Further Developments in Scientific Optical Imaging

Further Developments in Scientific Optical Imaging

Edited by

M. Bonner Denton
University of Arizona, Tucson, Arizona, USA

ROYAL SOCIETY OF CHEMISTRY

Chemistry Library

The proceedings of the International Conference on Scientific Optical Imaging held in Georgetown, Grand Cayman on 2–5 December 1998.

Special Publication No. 254

ISBN 0-85404-784-0

A catalogue record for this book is available from the British Library

Published by The Royal Society of Chemistry,
Thomas Graham House, Science Park, Milton Road,
Cambridge CB4 0WF, UK

For further information see our web site at www.rsc.org

Printed and bound by Bookcraft (Bath) Ltd

Preface

Solid-state imaging detector technology continues to revolutionize many disciplines. Astronomy and other aspects of low light level imaging, capture of low contrast images, and many areas of spectroscopy all benefit from the use of these devices. These devices offer high speed, sensitivity and dynamic range, making them desirable in a wide variety of applications, in industry, the military, general research, routine plant process control, and motion picture special effects.

Contained herein is a compilation of papers presented at the Fourth International Conference on Scientific Optical Imaging, held 2–5 December 1998, in Georgetown, Grand Cayman Island, British West Indies. These papers contain the latest information about commercial and academic research, development, and application in scientific optical imaging, from state-of-the-art devices to exciting explorations in space. Astronomers, spectroscopists, chemists, device and camera system developers and manufacturers, and optics specialists shared information about the projects in which they have been active.

The International Conference on Scientific Optical Imaging was organized for the purpose of providing a common ground for communication among diverse groups of scientists and engineers engaged in the design and application of optical array sensors. The conference program included lectures and posters presented by invited speakers. Plenary lectures given by world experts in their fields provided overviews of important aspects of optical imaging, such as design considerations, device fabrication and integration, and data reduction.

The program and venue were designed to place optimum emphasis on communication and cross-fertilization among disciplines. Lectures were held in the afternoons and evenings, leaving mornings free for informal discussion and interaction in the casual atmosphere of tropical Grand Cayman Island. Participants shared problems, solutions, and projected trends in this rapidly expanding field of endeavor. Developers, manufacturers, and users of this technology were able to interact in ways of mutual benefit and education to guide the development of new devices. Users from different fields exchanged ideas from their unique perspectives.

The editor would like to thank all participants for attending the conference and for sharing their knowledge and experience, particularly in the production of these papers. I would especially like to thank Ms. Christina Jarvis for her editorial assistance in preparing this volume, Ms. Julia Schaper for seeing to all arrangements for travel and accommodation, and David Jones and Jeffrey Giles, graduate student assistants, who provided able assistance to ensure that the conference was a great success.

Contents

ADVANCEMENTS IN SMALL-PIXEL, VIDEO-RATE, BACKSIDE-ILLUMINATED CHARGE-COUPLED DEVICES

John R. Tower

Sarnoff Corporation
CN 5300
Princeton, NJ 08540-5300

1 ABSTRACT

Sarnoff Corporation has developed a CMOS-CCD process technology that has enabled the development of a new generation of backside-illuminated charge-coupled devices (CCDs). The new devices achieve high quantum efficiency (QE), combined with high modulation transfer function (MTF) performance, at pixel sizes down to 6.6 microns. This new generation of CCDs also provides excellent noise performance at video readout rates. To permit the realization of large-format devices, photocomposition (stitching) has also been demonstrated within this new process technology.

2 INTRODUCTION

Sarnoff, formerly RCA Laboratories, began work on CCD development in 1971. The backside-illuminated CCD work at RCA dates back to 1978. The backside-illuminated process that has now been employed for over twenty years utilizes whole wafer thinning with backside lamination to a glass support substrate.[1, 2] The back surface is implanted and then furnace annealed to provide stable, high quantum efficiency from < 400 nm to > 1,000 nm.

Over the past few years, Sarnoff has been moving toward a more unified processing approach across our imaging products. Recently, we have integrated aspects of our double polysilicon spectroscopic CMOS-CCD process with our backside-illuminated, triple polysilicon CCD process. We have also moved the majority of our products to Canon 5X lithography to achieve tighter design rules. This paper will summarize the present state-of-the-art at Sarnoff.

3 PROCESS TECHNOLOGY

The new generation of backside-illuminated devices is processed in a triple polysilicon, single-level metal process technology with aluminum metallization. The process technology supports full CMOS circuitry in a twin well configuration. Standard oxide thicknesses are employed for pixel sizes > 12 microns, and scaled, thin oxides are employed for pixel sizes < 12 microns. The CMOS/CCD process supports CMOS-quality electrostatic discharge (ESD) pad protection.

A number of process options have been incorporated into the process flow. Pixel or horizontal register buried blooming drain structures can be implemented to achieve anti-blooming. Buried contacts can be realized to reduce the floating diffusion capacitance. This floating diffusion stray capacitance reduction increases the output sensitivity and reduces the amplifier equivalent noise.

To achieve high-speed clocking of vertical register gates, metal-to-poly strapping contacts have been demonstrated for pixel sizes down to 8 microns. These small geometry contacts are the enabling technique for achieving 1 MHz vertical clock rates on CCDs with > 50-mm-long gates at an 8-micron pixel pitch.

To produce very long linear array devices, Sarnoff has developed photocomposition (stitching) capability on the Canon 5X stepper. Employing stitching, Sarnoff has demonstrated > 60-mm-long devices. Figure 1 shows a long linear array produced with

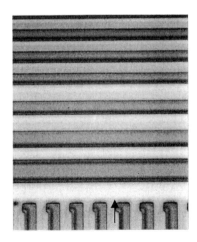

Figure 1 *Packaged linear array produced with stitching*

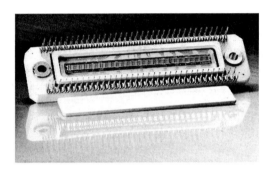

Figure 2 *Demonstration of stitching—panel-to-panel boundary indicated*

Table 1 *Measured Full Well Capacity*

Device	Pixel Size	Standard Process	Enhanced Process
Nova	8 μm × 8 μm	103,000 e	160,000 e
Mark V	6.6 μm × 6.6 μm	38,000 e	54,000 e

stitching. Figure 2 shows the excellent registration and feature size fidelity achieved with stitching. The figure indicates the stitch boundary location. The poly 2 gates at the top of the figure are 4 microns in width. The stitch boundary is difficult to detect, with typical panel-to-panel misalignments of 0.1 micron.

4 IMPROVEMENTS IN QUANTUM EFFICIENCY

The major advantage of backside illumination is the high quantum efficiency that can be achieved, particularly at wavelengths below 550 nm. With 100% optical fill factor and no gates to absorb short wavelength photons, backside-illuminated devices can approach ideal silicon quantum efficiency. Figure 3 indicates measured QE for the best non-AR coated devices peaking at 80%. With the AR coatings now being developed, the peak QE should be > 90%. Furthermore, as indicated in Figure 3, the AR coatings can be tailored to peak the QE in the band of interest.

5 IMPROVEMENTS IN DYNAMIC RANGE

As the pixel size is reduced, the dynamic range is compressed due to reduction in full well capacity. To maximize the dynamic range for small pixel devices we have 1) increased the area charge capacity ($e/\mu m^2$) of the pixel, and 2) decreased the amplifier noise floor. The measured improvement in full well will be discussed first. Two small pixel designs have been fabricated with standard buried channel implants and enhanced, high capacity implants. The measured full-well results are indicated in Table 1. The criteria for full well is the maximum charge that can be transferred without extended-edge-response charge-transfer efficiency (CTE) degradation. The Nova 8-micron pixel results indicate that > 1.5 X improvement in full well can be achieved moving from the standard implant to the high capacity implants. The charge density is > 9,000 $e/\mu m^2$ for

Table 2 *Measured High Sensitivity Output Amplifier Performance*

Specification	Measured Performance (23 C)
FD Sensitivity	27 μV/e
Output Sensitivity	15 μV/e
Output Non-Linearity	< 2% deviation
Output Noise (5 MHz clock)	8.9 e RMS
Output Noise (1 MHz clock)	5.8 e RMS

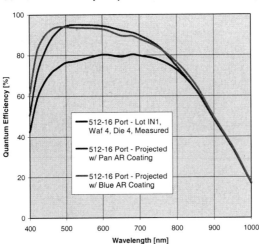

CCD Quantum Efficiency - Projected Performance with AR Coatings

Figure 3 *Measured QE and projected QE for backside illuminated CCDs*

the improved Nova device. The Mark V 6.6-micron pixel design does not exhibit charge densities as high as the Nova design. By layout changes, the design can be optimized further to provide higher charge capacity. However, as is, this charge capacity is excellent for such a small pixel device.

To achieve a full 12-bit dynamic range (> 4096:1) at video clock rates, a new generation of output amplifiers was developed for the small-pixel CCDs. The first of the new designs was a two-stage amplifier employing a buried contact.[3] The measured performance of the amplifier as demonstrated on the 6.6-micron pixel, 1k × 1k Mark V imager is shown in Table 2. The amplifier signal was processed with an off-chip correlated double sampling (CDS) analog processor for these measurements.

The output sensitivity of 15 μV/e is reflected back through the measured voltage gain of 0.54 V/V to a floating diffusion sensitivity of 27 μV/e. The predicted noise for cooled operation at −10C is < 7 e RMS at a 5 MHz clock rate and < 5 e RMS at a 1 MHz clock rate. The predicted maximum clock rate for this amplifier is 12 MHz, consistent with single port, 640 × 480 format, RS170 video rates.

6   IMPROVEMENTS IN MODULATION TRANSFER FUNCTION

A major challenge in achieving small-pixel, backside-illuminated CCDs is the attainment of high modulation transfer function (MTF) at practical silicon thicknesses. A rule of thumb has been that the silicon thickness should scale with the pixel size, with the pixel dimension no smaller than the silicon thickness dimension. The driver for this guideline was minimizing thermal diffusion of carriers outside the depletion edge for short wavelength illumination. Following this guideline the 6.6-micron pixel Mark V CCD would require a silicon thickness of roughly 6 microns to achieve reasonably high

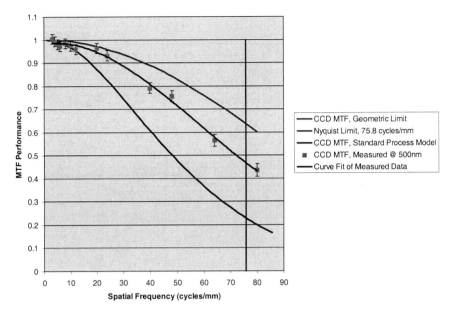

Figure 4 *Measured (adv. process) and predicted (std. process) MTF for 6.6-micron pixel CCD*

MTF. This silicon dimension is not practical from a manufacturing perspective. Furthermore, the thinner the silicon, the lower the broadband quantum efficiency.

Sarnoff has developed an approach to backside-illuminated imagers that provides high MTF for small-pixel devices at standard silicon thicknesses. Figure 4 shows the measured and modelled MTF for 500-nm illumination for the 6.6-micron pixel at a silicon thickness of 10 microns. At the Nyquist spatial frequency of 75.8 cycles/mm, the MTF is 46% (@ 500 nm). The graph indicates the Nyquist geometrical limit of 63% for the 100% fill-factor pixel. The graph also shows the predicted Nyquist performance of only 23% (@500 nm) for Sarnoff's historical, "large-pixel" process. The measured performance represents a 2X improvement in MTF compared with conventional backside-illuminated approaches.

Table 3 *Summary of Sarnoff's Process Technologies for Imaging Applications*

Process	Layers Polysilicon	Layers Metal	Pixel Technology
Spectroscopic CMOS/CCD	2	2	Virtual Gate Detector
Infrared CCD	2	1	PtSi Detector
Infrared CMOS	1	2	PtSi Detector
Visible CCD (>12-μm pixel)	3	1	Std.-backside CCD
Visible CCD (< 12-μm pixel)	3	1	Scaled-backside CCD
High-density CCD	4	3	Virtual Gate Detector

7 SUMMARY

Sarnoff has developed a new generation of small-pixel, video-rate, backside-illuminated CCDs. Devices have been demonstrated at 8-micron x 8-micron pixels and 6.6-micron x 6.6-micron pixels. This new generation of backside-illuminated CCDs complements our other imaging device technologies. Table 3 provides an updated summary of Sarnoff's process technologies for imaging applications.

References

1. P.A. Levine, D.J. Sauer, F-L. Hsueh, F.V. Shallcross, G.C. Taylor, G.M. Meray, and J.R. Tower, *SPIE Proc.*, 1994, **2172**, 100.

2. J.R. Tower, 'Application Specific Imager Design and Fabrication at Sarnoff', *International Conference on Scientific Optical Imaging Proc.*, 1995.

3. B.E. Burke, et. al., 'Soft-X-Ray CCD Imagers for AXAF', *IEEE Trans. Electron Devices*, October, 1997.

4. D.J. Sauer, F-L. Hsueh, F.V. Shallcross, G.M. Meray, P.A. Levine, G.W. Hughes, and J. Pellegrino, *SPIE Proc.*, 1990, **1291**, 174.

5. P.A. Bates, P.A. Levine, D.J. Sauer, F-L. Hsueh, F.V. Shallcross, R.K. Smeltzer, G.M. Meray, G.C. Taylor, and J.R. Tower, *SPIE Proc.*, 1995, **2415**, 43.

6. P.A. Levine, G.C. Taylor, F.V. Shallcross, J.R. Tower, W. Lawler, L. Harrison, D. Socker, and M. Marchywka, *SPIE Proc.*, 1993, **1952,** 48.

7. J. Ambrose, B. King, J. Tower, G. Hughes, P. Levine, T. Villani, B. Esposito, T. Davis, K. O'Mara, W. Sjursen, N. McCaffrey, and F. Pantuso, *SPIE Proc.*, 1995, **2552-37**, 364.

8. T.W. Barnard, M.I. Crockett, J.C. Ivaldi, P.L. Lundberg, D.A. Yates, P.A. Levine, and D.J. Sauer, *Analytical Chemistry*, 1993, **65**, 1231.

9. T.S. Villani, W.F. Kosonocky, F.V. Shallcross, J.V. Groppe, G.M. Meray, J.J. O'Neill III, and B.J. Esposito, *SPIE Proc.*, 1989, **1107-01**, 9.

10. T.S. Villani, B.J. Esposito, T.J. Pletcher, D.J. Sauer, P.A. Levine, F.V. Shallcross, G.M. Meray, and J.R. Tower, *SPIE Proc.*, 1994, **2225**, 2.

11. W. Kosonocky, G. Yang, C. Ye, R. Kabra, J. Lowrance, V. Mastrocolla, D. Long, F. Shallcross, and V. Patel, '360 × 360-Element Very High Burst-Frame Rate Image Sensor', IEEE International Solid State Circuits Conference, February 1996.

NEW DEVELOPMENTS IN THREE-DIMENSIONAL IMAGING WITH OPTICAL MICROSCOPES

Craig D. Mackay

PerkinElmer Life Sciences
Abberley House
Great Shelford, Cambridge, England

1 CONFOCAL IMAGING: INTRODUCTION

Optical microscopes are capable of extremely good depth resolution. A small change in the focal position can bring features sharply into focus. The performance of optical microscopes, however, is substantially degraded because of scattered light from other parts of the sample that are out of focus. In order to see how this happens, consider using not the normal uniform source of illumination but a single small pinhole illuminated from behind.

The objective of the microscope produces a converging beam that will produce a bright spot of illumination on one plane within the sample. If the eyepiece is also focused on the same plane where the hole is in focus, then the user will see a bright spot of light in the sample, surrounded by a halo of faint illumination. This halo arises because the column of light converging towards the bright spot is also illuminating the sample with its out-of-focus beam on either side of the focal plane. The level of illumination in out-of-focus planes is considerably smaller than that found in the focussed spot, but nevertheless the general low-level halo from the spot is significant. When a sample is illuminated with a continuous uniform field, then every element that is illuminated also gives rise to a similar halo around it and what the user sees is a summation of all the illuminated spots and their halos. For this reason, when a conventional microscope is focused on a specific plane within the sample, the background scattered light level can be very high indeed.

One elegant solution to the problem is to use the confocal arrangement. Think again about the case where a single pinhole illuminates the sample. The optics can be arranged so that only light actually emitted by the brightly illuminated spot will be passed onto the detector system, and faint halo light will be suppressed by a second pinhole, also in a plane focussed on the illuminated spot. This is the confocal arrangement. The most common arrangement for confocal microscopy is to generate a bright spot of light by using a scanning laser beam, ensuring that the laser beam illumination spot is always confocal with a pinhole scanned in synchronism with the laser. This way it is possible to suppress very greatly indeed the out-of-focus illumination to give remarkably clear and sharp pictures. The disadvantage of this arrangement is that only one element of the sample is illuminated at once, and obtaining an output image of reasonably good resolution may require many seconds of scanning to create a single image. A further difficulty is that in order to get an adequately strong signal, the laser beam must be very bright, which can often lead to photobleaching with fluorescence microscopy.

2 HIGH-SPEED CONFOCAL SOLUTIONS

There are many applications in optical microscopy where it is important to be able to take many images of the sample per second. As we have seen above, with a conventional scanning laser microscope it is not possible to do this without using exceptionally high light levels. An elegant solution to this problem is to employ a white light source, rather than a laser beam, to illuminate a mask comprising a large number of pinholes. The confocal mask also has pinholes, which are scanned in synchronism with the illuminated mask. A large number of pinholes may therefore be scanned across the sample at once, reducing the scanning time dramatically and also allowing confocal microscopy to be carried out with non-laser sources such as tungsten lamps, with appropriate colour filters as required. It also allows genuine white light confocal imaging, which may be convenient in applications where the sample has its own colour. This scanning disk, known as a Nipkow disk, has been used for a number of years to provide high-speed confocal imaging. The disadvantage of the multiple pinhole approach is that in order to ensure that the general background illumination does not get too high, the pinholes must be well spaced and therefore the overall light transfer efficiency through the pinhole mask becomes relatively poor. Tungsten lamps are not able to produce the higher brightness obtainable with the laser, and therefore Nipkow disk confocal microscopes suffer from relatively low sensitivity, particularly in applications such as fluorescence microscopy.

3 FAST, EFFICIENT REAL-TIME CONFOCAL IMAGING

A more recent development by research workers in Japan has been commercialised by Yokogawa. In their system, the overall light gathering efficiency of the Nipkow disk is improved by providing a second disk before the pinhole disk, consisting of a large number of tiny lenslets, each of which is focused on a corresponding pinhole. The amount of light that passes through the pinhole is greatly increased because each tiny lens is able to capture the illuminating light across the aperture of the lenslet rather than simply across the aperture of the pinhole. The lenslets and the pinholes are arranged in a spiral pattern and the disk is rotated to give approximately 360 images per second.

Recently, a commercial system has been developed by PerkinElmer Life Sciences and is now being sold by them worldwide. The system is integrated with a high-speed digital CCD camera and a multiple-wavelength laser. Images may be obtained very rapidly indeed. One viable mode of operation is to take an image at one focus setting and then to step the stage in the same direction by a small amount and repeat the image capture. This procedure is repeated many times, effectively building up three-dimensional data cubes of images, allowing genuine three-dimensional confocal image data sets to be acquired in real time. The system presently marketed by PerkinElmer Life Sciences uses these data sets with a sophisticated software package to generate 3D images of the sample. These images may be displayed on a computer screen and rotated and zoomed in three dimensions so that the microscopist may inspect the sample in whichever orientation and in whichever way is most appropriate.

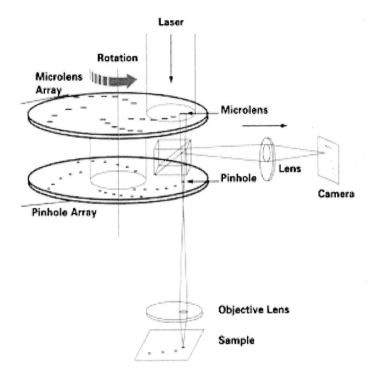

Figure 1 *The arrangement of scanning lenslets used in the Yokogawa head, which is part of the PerkinElmer Life Sciences High-Speed Confocal Imaging System*

4 DIGITAL (NON-CONFOCAL) DECONVOLUTION

There is another approach that allows one to record three-dimensional images without the complication of using a confocal microscope. Images are taken at a set of stepped focal positions in the sample, just as with the fast scanning confocal microscope described earlier. As previously explained, the difficulty here is that each image contains the desired in-focus image plus a considerable contribution of light from the out-of-focus images. It is now possible to purchase powerful computers able to sort out in-focus from out-of-focus images. This process is known as deconvolution. In principle, what it does is as follows. The programme looks in one of the image planes for a feature that is very compact and is essentially unresolved with the optical configuration used. By having good knowledge of the system's optical configuration (and a detailed description is essential for the system to work as well as it can), the software is able to compute the contribution from the other image planes from this element in the one plane that is in focus. It then subtracts the out-of-focus component corresponding to that single bright spot for each of the other image planes. This procedure is repeated for every bright element in the three-dimensional image data sets and the process is continued and repeated again and again until what remains is a substantially confocal image.

This method has the advantage over confocal microscopy of allowing conventional microscope systems to be used and particularly of having a very high light efficiency. The difficulty with this method is that in situations where the sample to be imaged is relatively weak and the out-of-focus images are relatively bright, the overall signal to noise that may be achieved is compromised in any one plane by noise contributions still present from other image planes. The process is in essence a complicated way of subtracting one image plane from another adjacent (and therefore similar) one and looking at the differences. If the signals are large and the differences are small images, it will be clear that the overall signal-to-noise must be compromised.

In practice the choice between digital deconvolution and scanning confocal microscopy is very much dependent on the kind of sample used. The software system provided by PerkinElmer Life Sciences with the scanning confocal system described in the last section also has software facilities that allow the digital deconvolution method to be used, allowing the customer to choose the method that happens to best suit his or her application.

5 FAST THREE-DIMENSIONAL RECONSTRUCTION METHODS

One of the principle limits to a wider acceptance and use of digital deconvolution is the substantial amount of computer processing time needed to make a good job of working out exactly what the three-dimensional images are. Deconvolution methods such as maximum entropy deconvolution, one of the methods most favoured for this kind of work, is limited because of the mathematics involved in getting the computation to converge. In its standard form, the maximum entropy method may only be used for functions that are strictly non-negative. The mathematical consequence of this is that the reconstruction methods that must be used with maximum entropy (and with most other deconvolution methods) must be done in Fourier space using Fourier transforms. This holds even though for most applications it will be much more appropriate to use an entirely different base for transformation and reconstruction calculations.

PerkinElmer Life Sciences has recently patented a different mathematical procedure, developed to overcome this limitation, which allows a more appropriate basis to be used. The effect is that in virtually every case it is possible to speed up the calculation of these multi-dimensional data sets by a factor in excess of 50. This means that it is now possible to carry out maximum entropy deconvolutions on much larger data sets on relatively moderate-performance personal computers at speeds substantially faster than has been practical for even the most powerful Unix workstations.

These deconvolution methods have also been patented more widely, as they have applications in many other areas, including geophysics, radar and sonar pattern analysis, and many other kinds of image processing and data stream reconstruction.

MEGACAM: A WIDE-FIELD IMAGER FOR THE MMT OBSERVATORY

B. A. McLeod, M. Conroy, T. M. Gauron, J. C. Geary, and M. P. Ordway

Harvard-Smithsonian Center for Astrophysics
60 Garden Street
Cambridge, MA 02138 USA

1 INTRODUCTION

The Multiple Mirror Telescope (MMT) on Mt. Hopkins, Arizona, is currently being converted from six 1.8 m telescopes to a single 6.5-m-diameter mirror. The imaging configuration of the f/5 wide-field-corrected Cassegrain focus will provide a 35'-diameter field, which is flat and has 0".1 RMS image quality from 0.35 to 1.0 μm. Megacam will populate this focal plane with state-of-the-art CCD detectors. The mosaic will consist of 18432 × 18432 0".08 pixels. We have previously described a design for this camera.[1,2] In this paper we provide an updated description of the planned hardware.

2 CAMERA HOUSING

The camera housing, or "topbox", is a 2-m-diameter, 0.3-m-deep steel structure that contains the shutter and filter wheels, and provides a rigid interface between the CCD Dewar and the telescope. A cut-away view of the assembly is shown in Figure 1. When not on the telescope, the housing is mounted on a custom cart with a trunnion to allow easy rotation. For storage and handling, the instrument is rotated 90° to take up minimum floor space.

2.1 Filter Wheels

The housing contains two overlapping filter wheels, each with five slots. Normally, one of these slots will be left blank in each filter wheel, allowing a choice of eight filters at any given time. The filter wheels are driven from their centers with a DC servo motor, harmonic drive gear reduction, and a rotary encoder, which provides a lateral repeatability of 10 μm at the radius of the filters. This high repeatability allows us the option to use segmented filters with masks between the segments. Without high positioning repeatability, flat-fielding errors would occur as the mask shadow shifts on the detector. Each filter wheel is composed of two steel facesheets separated by a web of reinforcements. Filters are mounted in a steel frame that is inserted between the two facesheets of the filter wheel from access ports on each side of the housing. During

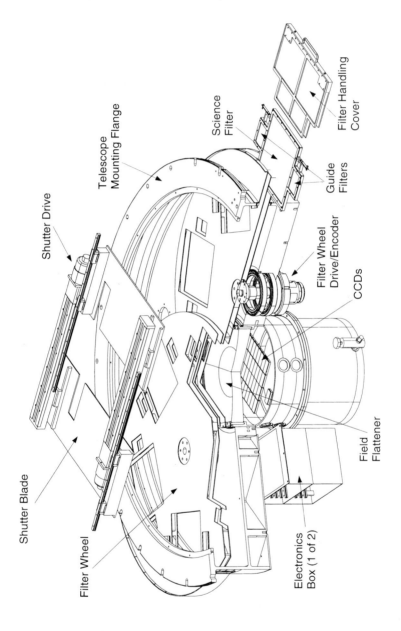

Figure 1 *Cutaway and exploded view of Megacam*

Shutter Drive

Telescope
Mounting Flange

Science
Filter

Filter Handling
Cover

Guide
Filters

Filter Wheel
Drive/Encoder

CCDs

Shutter Blade

Field
Flattener

Filter Wheel

Electronics
Box (1 of 2)

insertion and removal, the filter is completely enclosed by an aluminum handling cover to minimize risk.

2.2 Shutter

The shutter consists of the two aluminum plates mounted on pairs of THK rails. Notches are cut in the shutter blades so that light reaches the guide chips when the shutter is closed. It is also possible to close the shutter completely so that no light hits the guide chips. The exposure is taken by opening one blade and ended by closing the other blade. This ensures an even exposure time over the full focal plane. A shutter blade traverses the focal plane in less than 5 s; however, shorter exposures can be achieved by having the closing blade follow the leading blade closely, effectively scanning a slit across the focal plane. This style of shutter is used in the Big Throughput Camera.[3] The shutter is driven from one edge with a DC servo motor and a lead-screw with a rotary encoder. The blade is attached to the other rail with a flexure to allow for differential expansion between the steel structure and the aluminum shutter blade without binding.

3 MEGACAM DEWAR

3.1 CCD Package

We plan to use EEV CCD42-90 model thinned back-illuminated CCDs with 2048×4608 13.5-μm pixels. We have taken delivery of an initial order of six devices. Two of these devices are being used in a prototype camera called Minicam. The other four will be used in the Hectospec and Hectochelle fiber spectrographs at the converted MMT.[4,5] These CCDs have a custom designed package to allow four-edge butting (see Figure 2). Signals from the silicon are wire-bonded onto a ceramic header that wraps around to the bottom of the Invar package, where the signals are accessed from a pin grid array. A thermistor potted into the Invar is also connected to the pin grid array. Electrical connections to the CCDs are made via a custom zero-insertion-force (ZIF) socket.

The CCDs will be mounted on a 6-mm-thick Invar plate that is ground flat. This plate is thermally isolated from the Dewar case with a set of six titanium flexures mounted around the edge of the plate. The flexures allow for the 125° C temperature differential without bending the plate, which is critical for keeping all parts of the large focal plane in focus. When mounted in the mosaic, the gaps between the imaging areas will be 1 mm (corresponding to 6") on three sides, and 6 mm (36") on the fourth side.

Without special tools there is no easy way to pick up four-edge buttable CCDs for installation. Thus, we handle it only via a rod that screws into the back. The handling rod, along with another temporary guide rod and two permanently attached locating pins, prevents the CCD from touching its neighbors as it is inserted into the focal plane. With practice, handling these CCDs is quite straightforward. We are making no attempt to align the CCDs to each other to better than a few pixels, because the camera will not be used for drift scanning. The relative orientations will be determined by observing star clusters with known astrometry. The coplanarity of the CCDs in the mosaic is achieved

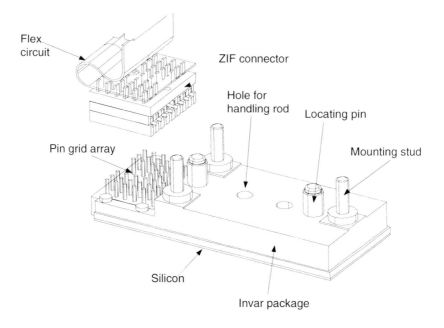

Figure 2 *CCD42-90 connected to flex circuit*

by a set of precision shims attached to the Invar package by the manufacturer. These shims also provide the thermal contact between the CCD and the cold mounting plate.

3.2 Cooling

Cooling a large CCD mosaic provides a number of challenges. The focal plane must be kept a uniform temperature, electrical cables must pass to the middle of the mosaic, and, at the same time, easy access to the CCDs must be preserved for maintenance and upgrades. Because Invar is a relatively poor thermal conductor, the cooling path must be distributed evenly over the Invar plate. Initially, we considered using a radiative cooling mechanism,[2] but have concluded that such a system would be only marginally adequate. Instead, we are adopting a conductive system. As seen in Figure 3, a copper cold distribution plate is mounted above the Invar plate, and the two are connected by a set of straps, one for each CCD. Flexible electrical cables going to each CCD pass through the distribution plate. The distribution plate will be cooled with four Cryotiger closed cycle coolers. Alternatively, a pair of liquid nitrogen Dewars could be used instead. This geometry allows one to adjust the cooling straps without disconnecting the CCDs, and allows CCD changes to be made without disconnecting the cooling straps or the Cryotigers. Access to the back of the focal plane is gained by removing a single panel from the Dewar.

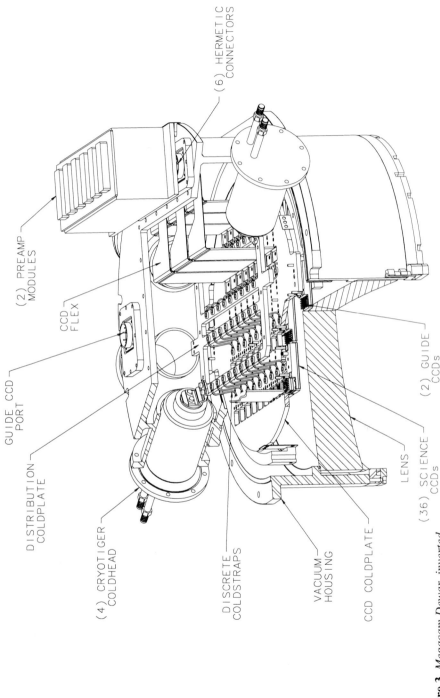

Figure 3 *Megacam Dewar, inverted*

4 ELECTRONICS

Here, we provide an overview of the signal chain from the CCD to the computer. Details of the electronics are described in an accompanying paper in this volume.[6] Interconnections between the various elements rely heavily on custom flexible printed circuits to facilitate assembly and increase reliability.

Sets of three CCDs are connected via the custom ZIF connectors to a flexible circuit that is soldered to a hermetic connector on the wall of the Dewar. We will solder a small amount of passive components to the warm end of the flex cable to provide filtering and overvoltage protection.

Immediately outside the Dewar will be a preamp module that mates to the hermetic connectors on one side and has a set of second custom flexible circuits on the other. These flex cables lead to the backplane of the main electronics chassis. The flex circuits will be made with several layers to allow for shielding of the video lines.

The electronics are divided into two boxes mounted on either side of the Dewar. Each of the two boxes will be responsible for driving signals for eighteen CCDs and digitizing the data from their 36 output amplifiers before bussing the data back to the master box for transmission to the host computer. Analog levels on the driver boards are set with D/A converters individually for each CCD. Each driver board can control three CCDs. Each signal processing board can service four channels, i.e. two CCDs.

In addition to the driver and video boards, one of the boxes, the master, contains two additional boards for timing generation and communications. On the timing board, clock signals are generated in digital form and transmitted via the back plane to the analog clock driver boards. The clock signals are generated from waveforms and sequences stored in RAM chips on the timing board that are loaded from the host computer. The CCDs will be clocked at 200 kilopixels s^{-1}, yielding a readout time of 24 s at full resolution, or 3 s if the pixels are binned 3×3 to 0".24. The logic circuitry on all boards is largely consolidated into Altera programmable gate arrays. There is no microprocessor in the CCD electronics.

Communication with the host computer for programming the controller and receiving data is done with the EDT PCI-RCI interface system. The camera module of the interface is mounted on a board in the master box. Programming commands to the electronics are transmitted over an optical fiber on a 110 Kbaud channel. The data channel will support the full 29 Mbyte s^{-1} rate of Megacam over a single optical fiber link. The computer end of the PCI-RCI system resides in a Sun Sparc Ultra-60 running Solaris 2.6. This system is adequate for the Minicam and spectrograph applications, but we will upgrade to a faster machine before the full Megacam focal plane is complete.

For housekeeping, all analog voltages and CCD thermistors are multiplexed onto a Keithley Smartlink multimeter for monitoring. Data from the Smartlink is read over the local ethernet. With this configuration, the CCD electronics on the telescope are connected to the control room with only a network connection and the EDT link.

References

1. B. A. McLeod, D. G. Fabricant, J. C. Geary, and A. H. Szentgyorgyi, in 'Solid State Sensor Arrays and CCD Cameras', C. N. Anagnostopoulos, M. M. Blouke, and M. P. Lesser, eds., *Proc. SPIE*, 1996, **2654**, 233–238.

2. B. A. McLeod, M. Conroy, T. M. Gauron, J. C. Geary, and M. P. Ordwayin 'Optical Astronomical Instrumentation', S. D'Odorico, ed., *Proc. SPIE*, 1998, **3355**, 477–486.

3. D. Wittman, J. A. Tyson, G. M. Bernstein, D. R. Smith, R. W. Lee, M. M. Blouke, I. P. Dell'Antonio, and P. Fischer, in 'Optical Astronomical Instrumentation', S. D'Odorico, ed., *Proc. SPIE*, 1998, **3355**, 626–634.

4. D. G. Fabricant, E. H. Hertz, A. H. Szentgyorgyi, R. G. Fata, J. B. Roll, and J. Zajac, in in 'Optical Astronomical Instrumentation', S. D'Odorico, ed., *Proc. SPIE*, 1998, **3355**, 285–296.

5. A. H. Szentgyorgyi, L. W. Hartmann, P. N. Cheimets, D. G. Fabricant, M. R. Pieri, and Y. Zhou, in 'Optical Astronomical Instrumentation', S. D'Odorico, ed., *Proc. SPIE*, 1998, **3355**, 242–249.

6. J. C. Geary, in 'Further Developments in Scientific Optical Imaging', M. B. Denton, ed., Royal Society of Chemistry, Cambridge, 2000, 18–23.

SIGNAL PROCESSING FOR THE S.A.O. MEGACAM

John C. Geary

Smithsonian Astrophysical Observatory
60 Garden St.
Cambridge, MA 02138

1 INTRODUCTION

The SAO Megacam[1] will be a massively parallel 72-channel CCD array camera, with the main focal plane holding 36 identical 2-channel imagers, of format 2K × 4.5K. We intend to read this large structure out at a pixel rate of 200 kHz, thus giving a total readout time of approximately 11 s for an unbinned image and correspondingly less for the binned images that will be used most of the time. Because this readout rate is several times what we have used in the past, a program to optimize the CCD signal processing chain was undertaken and has now been successfully implemented into the first prototype versions of the Megacam controller.

2 THE SAO PRE-AMPLIFIER

In a previous study[2], we investigated the use of individual IC circuits for CCD pre-amplifiers. At the modest readout rates of an earlier generation of devices, typically less than 100 kHz, it was found that quite acceptable noise performance could be obtained from the better FET-input op-amps, such as the AD745 and OPA627. However, it was noted then that trouble would occur at some point if faster readout was desired, as the settling time for both these amplifiers is excessively long for fast work. This is especially true if one wants to retain a dual-slope integrator scheme for post-pre-amp signal processing in order to optimize noise performance of the system. Under these circumstances, the long settling tails of the best FET op-amps intrude into the desired integration intervals when faster readout is attempted.

To find much faster amplifiers with short, well-behaved settling characteristics, one is forced at present toward some of the newer low-noise bipolar-input op-amps. However, in order to present high impedance input to the CCD output, we are obliged to mate a discrete low-noise FET to this op-amp as a front end. While this could be done as a common-source FET configuration, having front-end FET voltage gain augmented by the following op-amp, such a configuration tends to be phase-unstable and will usually break into high frequency oscillation unless inferior parts with peculiar phase shifts are used. A much more stable configuration is to use the input stage FET in a source-follower configuration, with all of the closed-loop gain then being supplied by the following op-amp stage, which

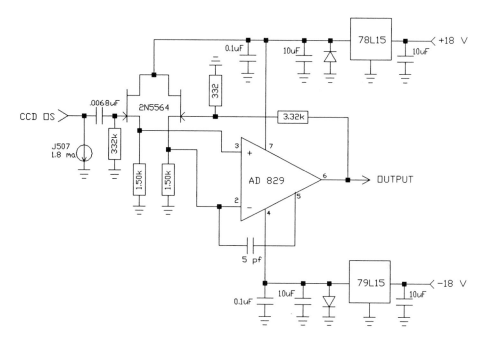

Figure 1 *The S.A.O. pre-amplifier, shown with a voltage gain of 11. Local power regulation is done to decrease crosstalk in multiple channels. Settling time is quite sensitive to the small 5 pf current feedback capacitor.*

must of course be low noise. The schematic for the SAO pre-amplifier is shown in Figure 1.

The choice of input FET is dictated more by a lack of any component development than by anything else. There have been few new discrete transistor products developed in recent years, as manufacturers concentrate wholly on ICs for mass markets. Indeed, it is often difficult to even find some favorite transistors of past years, as distributors purge them from their inventories. We have been using the 2N5564 dual matched JFET with good success for many years, and these transistors are usually still available through normal distribution channels. They have noise levels below 2 nV/rootHz at high frequencies and good 1/f characteristics. There are a few other dual JFETs that could be used, but not very many, and sometimes they are not readily available.

Choosing a bipolar op-amp for this design is more interesting, as several new, fast, low-noise products have been introduced in recent years. If, however, you wish to work with full +/-15V rails, your choices will be more limited, as many of the newer devices were developed for much reduced voltage levels. Because we wished to have the increased dynamic range of the higher rails, we chose for the op-amp the fairly recent AD829. This amplifier has a slew rate up to 200 V/usec, settling time to 0.1% of just 90 ns, and a flat noise voltage of about 1.5 nV/rootHz from 100 Hz onward. It seemed like an excellent candidate for a fast-settling pre-amp and soon proved to have the right stuff with minimum tweaking. Of particular interest in optimizing settling time is the current feedback option provided in this package, allowing one to tune the high frequency response and damping

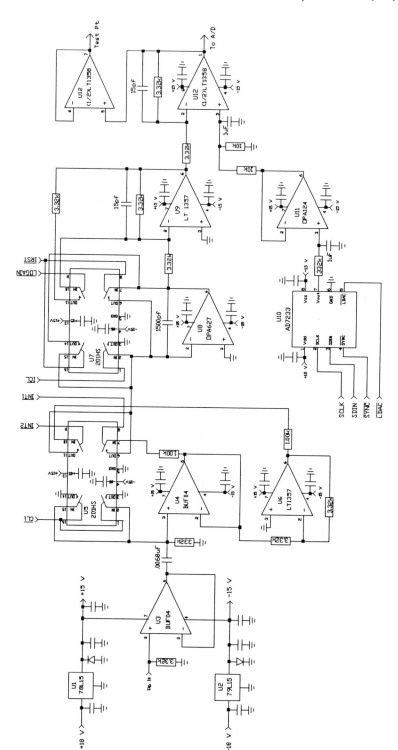

Figure 2 *CDS signal processor for a single Megacam channel. A X2 gain select is provided as well as digitally controlled DC offset and local power regulation.*

through the use of one small capacitor. Empirically adjusting this feedback capacitor for CCD inputs has allowed settling times close to 100 ns (0.1%) to be realized for full power 10 V output swings. This is almost an order of magnitude better than any FET amplifier we have tested, with comparable noise performance.

3 CDS SIGNAL PROCESSOR

For many years, we have used in our CCD controller, with good effect, a combination of dual-slope integrator with input AC clamping. While a dual-slope integrator alone is capable of providing the analog subtraction necessary to implement correlated double-sampling (CDS) signal processing, the addition of a clamped input helps to provide an additional level of baseline restoration and thus aids in combating nonlinearity effects, especially at high signal levels. To achieve the higher 200 kHz pixel rates, however, required some laboratory work to characterize a variety of candidate amplifiers that would provide the necessary performance in terms of both speed and noise. The results of this work have led us to the design shown in Figure 2.

In our input stage, we would have liked to have used a fast FET buffer for the clamped stage (U4), but we have been unable to locate a modern, low-power candidate for this job. The old LH0033 series of devices are fast enough, but run exceedingly hot and are not available in any miniature package. Because there is a lack of more modern FET buffers to choose from, we investigated some of the bipolar buffers that have been developed and found that the input impedances were high enough for use in this clamped stage. Both the BUF04 from Analog Devices and the BUF634 from Burr-Brown have been tried with good effect here. The latter device also has enough output current capability to be used as driver for the CCD itself in another part of the camera controller. The switches used for the clamp, as well as for controlling the dual-slope integrator, are all 201HS quad packages. These switches are the fastest available at the present time for +/-15V power rails and have an on-resistance below 50 ohms. Somewhat better performance can be had from some of the reduced-voltage parts that have been recently introduced, if one is designing for a +/-5V system.

The integrator stage utilizes the OPA627, which is an older part but which has not been bettered in terms of noise and slew rate, the latter a necessary specification for a fast integrator. Extensive use is made of the low-noise fast amplifiers from Linear Technology for most of the other parts of the signal chain. The ones chosen for this design have proved to be well behaved, and even faster versions are available if higher pixel rates are intended.

4 THE DIGITIZER AND BUS

The digital output stage for each of the four analog channels per VME signal processing board is shown in Figure 3. It was initially hoped that the recently introduced LTC1604 A/D converter from Linear Technology could be used in this design. It comes in an extremely small package and uses very little power, thus allowing more analog channels per board in a massive system such as the full Megacam. In order to evaluate its performance, a full 4-channel prototype system was built up for the converter and tested on an actual EEV 42-90 device. The results showed that while most aspects of this device were acceptable, the rather bad code-dependent differential nonlinearity probably was not.

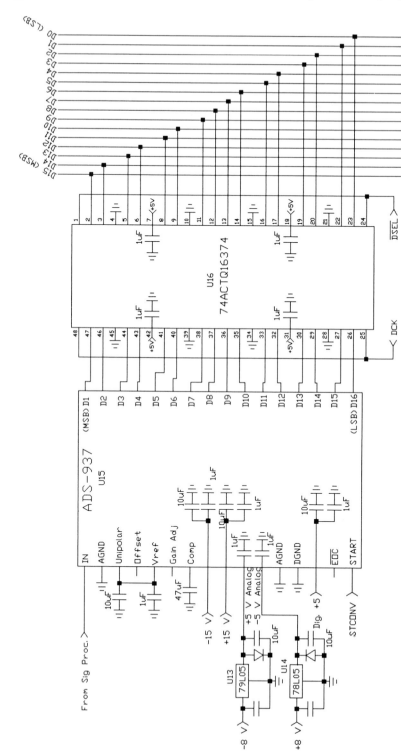

Figure 3 *A/D converter and digital bus interface. Local power regulation is provided for analog voltages. The digital bus is controlled by sequentially enabling the 3-state outputs of each channel in the system following latching of the parallel output data.*

Despite extensive efforts to ameliorate this difficulty through differing ground and power supply schemes, the non-linearity remains a problem and will probably relegate this prototype to use for guiding cameras only.

The converter we intend to use on the next version of the signal processor is the Datel ADS-937, a very fast part that has been extensively tested in our prototype camera and shown to be trouble-free. Its drawbacks are its size, power requirements, and cost, but there is no other available converter at present that does as well, overall, in these respects. The parallel output from each of the four channels per board is buffered from the common 16-bit data bus by a fast 3-state latch. This board bus is further buffered from the backplane bus by 3-state drivers, thus allowing the large number of digital channels of the full Megacam to be multiplexed into the I/O channel at the required rate.

References

1. B. McLeod, T. Gauron, J. Geary, M. Ordway, and J. Roll, *Proc. SPIE*, 1988, **3355**, 477.
2. J. Geary, *Optical Detectors for Astronomy*, J. Beletic and P. Amico (eds.), 1998, 127–130.

ADVANCES IN SCIENTIFIC-QUALITY DETECTORS AT JPL: HYBRID IMAGING TECHNOLOGY

Mark Wadsworth, Tom Elliott and Gene Atlas*

California Institute of Technology / Jet Propulsion Laboratory, Mail Stop 300-315, 4800 Oak Grove Drive Pasadena, CA 91109

*Adept IC Design, 7720 El Camino Real, Suite 2D Carlsbad, CA 92009

1 ABSTRACT

Recently work was begun at JPL to create a next-generation imager technology, called Hybrid Imaging Technology (HIT), that offers scientific-quality imaging performance. The key principle of this technique is the merging of CCD and CMOS technologies by device hybridization rather than by process integration. HIT offers the exceptional quantum efficiency, fill factor, broad spectral response, and very low noise properties of CCD imagers with the low-power operation and flexibility of integration found in CMOS devices. In this work we present the architecture, benefits, performance, and future directions of HIT.

2 INTRODUCTION

The field of solid-state image sensing has seen many new and exciting developments in the past several years. A frenzied industry push for the development and commercialization of CMOS imager technology for the consumer camera market motivated CCD manufacturers to mature their own fabrication processes, thereby improving yields and reducing costs. This focus has resulted in a myriad of low-cost CCD and CMOS imagers suitable for low-to medium-end consumer uses. While such devices offer an excellent starting point from which to develop scientific quality imaging chips, heightened industrial competitiveness has increased the reticence of many CCD and CMOS imager manufacturers to optimize existing commercial imager designs for the higher performance required by the comparatively small community of scientific users.

A similar situation exists in the arena of integrated circuit processing. Manufacturers and foundry services that previously supported flexible fabrication processes to accommodate niche markets have recently, for economic reasons, become disinterested in developing and maintaining non-standard process flows. Process variants that offer promise for substantial improvement in the performance of imaging devices have either become unavailable or far too expensive to develop and implement. In short, industry continues to increase and improve its integrated circuit fabrication capabilities even as some in the imaging community find it more difficult to exploit these capabilities for scientific purposes.

In 1997 work was begun at JPL to create a next-generation imager technology called Hybrid Imaging Technology (HIT). The goal of this development effort was the creation of a next-generation scientific-quality imaging approach that could utilize the latest advances in both CCD and CMOS fabrication technologies without process development or integration. The following sections describe initial results of this effort, including basic HIT architecture, performance of a pathfinder device, and suggestions for future improvements in the approach.

3 HIT ARCHITECTURE

3.1 Detector Configuration

The key advantage of HIT is the merging of CCD and CMOS technologies at the component level after, rather than during, device fabrication. Figure 1 compares existing CCD technology with the recently demonstrated HIT concept and indicates the supplementary steps required to incorporate additional functions into a HIT detector.

The imaging portion of a HIT detector is identical to that found in a CCD array. Charge is collected and held in a series of pixelated wells formed in an epitaxial silicon layer by electrically isolated, optically transparent, polysilicon control gates that overlay the imaging area. Collected charge is transferred in a pixel-by-pixel manner to the output structure by the application of a sequence of pulses to the control gates. However, unlike a CCD, the charge-to-voltage conversion in a HIT detector takes place in a charge-mode amplifier constructed in CMOS technology. This CMOS amplifier is bump bonded to the main imaging portion of the array using standard technology available in industry.

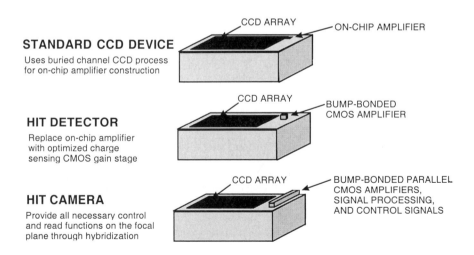

STANDARD CCD DEVICE
Uses buried channel CCD process for on-chip amplifier construction

HIT DETECTOR
Replace on-chip amplifier with optimized charge sensing CMOS gain stage

HIT CAMERA
Provide all necessary control and read functions on the focal plane through hybridization

Figure 1 *Hybrid Imaging Technology merges CCD imaging technology, CMOS signal processing technology, and infrared detector packaging technology to create a new class of ultra-high performance imagers for scientific applications*

Because the imaging portion of the HIT detector is simply a CCD, the resulting imager possesses all of the strengths of CCD technology, i.e., the HIT approach provides an inherent 100% optical fill factor and excellent quantum efficiency over a very broad spectral range (soft x-ray to near-infrared). It is also completely compatible with backside thinning and illumination. These benefits are supported only by HIT (and its parent CCD) technology, and cannot be implemented in charge-injection device (CID), active pixel sensor (APS), or any other standard CMOS-based device due to inherent architecture and CMOS fabrication process-related problems.

Device hybridization allows both technologies to be independently optimized, providing the ultimate imaging performance in a highly miniaturized format. Hybridization also allows for reuse of imager arrays and CMOS readouts without the costly process of refabrication. A supply of unhybridized components can be maintained and appropriate combinations of components can be selected to suit specific applications. Thus, HIT is a robust, reconfigurable technology capable of supporting a wide variety of scientific imaging needs.

3.2 CMOS Amplifier

The principal weakness of CCD technology is the on-chip amplifier. CCDs use buried channel process technology that, while optimized for low-frequency noise performance, has many undesirable features related to imager formation and charge transfer requirements. These retarding elements are absent in CMOS processing. HIT uses a charge-based gain stage similar to an operational amplifier configured as a charge integrator. The design of this CMOS amplifier has potential for lower noise operation during charge detection than that of either the typical on-chip buried channel amplifier found on a CCD or the CMOS-based amplifier found in CID or APS imagers for an identical data bandwidth. The HIT advantage over the CCD-based amplifier is one of process. However, HIT can also outperform CID and APS amplifiers, as there are no significant size constraints for the HIT amplifier (the HIT amplifier does not need to be squeezed onto the imaging chip). Thus the HIT amplifier can optimize transistor sizes to obtain the optimum 1/f and thermal noise performance. In addition, the HIT amplifier has roughly the same very low power utilization found in CMOS imagers, which is typically an order of magnitude lower than that of a CCD-based amplifier, yielding a dramatic reduction in overall device power consumption.

3.3 System Integration

Other components can be added to the CMOS portion that provide for full signal processing capability, including correlated double sampling and on-chip analog-to-digital conversion. Clock timing and driver circuitry can also be incorporated, creating a full camera system in a single integrated circuit package. This microcamera system offers the excellent imaging performance of a scientific-quality CCD camera with just slightly more power utilization than an APS single-chip camera[1]. In addition, as compared to a CCD-based camera, the mass of the HIT camera could be reduced by a full order of magnitude and would typically be dominated by the mass of the optics.

4 PROTOTYPE HIT DETECTOR

The prototype HIT detector is a 256 × 512 element array design with a 12-micron-square pixel. The imaging array was designed at JPL and fabricated using standard buried-channel CCD technology at Lockheed Martin Fairchild Systems (LMFS). A companion CMOS charge-mode amplifier was designed at Adept IC Design using the Hewlett Packard 3.3 volt, 0.5 μm, CMOS process and fabricated through the MOSIS foundry service. To assist in imager screening, both a standard buried-channel CCD amplifier and a HIT output node were incorporated, at opposite ends, into the serial register design. Initially, the standard on-chip amplifier was used to verify the imager design and functionality. Similarly the CMOS design was verified by packaging an individual chip and illuminating the device with visible light. Both imager and amplifier designs were found to function as designed. After functional verification, the CCD and CMOS components were hybridized at LMFS. Figure 2 is a photomicrograph of a prototype 256 × 512 HIT after hybridization and packaging.

For the proof-of-principle HIT device, the 2-mm-square CMOS chip contains an operational-amplifier-based charge integration amplifier connected by indium bumps to the underlying imager array. A total of nine bump interconnects were necessary to provide signal, control, and power inputs to the CMOS. All CMOS inputs are connected to the outside world by means of standard bond pads on the imager substrate. Only roughly 25% of the 2-mm × 2-mm area was required for the amplifier and bump interconnections. The remainder of the chip contains evaluation transistors and test circuits that provide information to allow optimization of future versions of the CMOS amplifier.

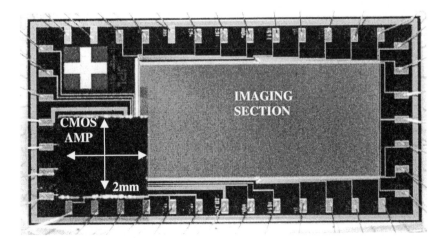

Figure 2 *A prototype 256 × 512 HIT detector consisting of a CCD imaging portion and a CMOS charge-detection amplifier*

5 DETECTOR CHARACTERISTICS

HIT detector characteristics were measured at JPL using a standard CCD test station. Test parameters resembled those of a typical CCD. Measurements, including imaging, linearity, and photon transfer characterization, were taken over a range of pixel rates from 12.5 kHz to 100 kHz and at device temperatures spanning -40 C to +30 C. The HIT device displayed image quality and noise performance rivaling that of scientific-grade CCDs while dissipating much less power. Figure 3a is an image obtained at 25-kilopixel/sec data rates in which the brightest portion of the scene, the neckerchief, corresponds to 50 electrons. This image has not been corrected for dark current nonuniformity and has not been processed or enhanced in any way. Operating at a temperature of -40 C, the detector produces minimal bright pixels and exhibits charge-transfer efficiency greater than 0.99999. The rms noise floor of the HIT detector at the 25-kilopixels/second rate is 4.8 electrons, and the total detector power dissipation is less than 100 μW. Figure 3b shows the detector performance at a 3000-electron image level when operating at 0 C.

A parameter of primary interest in any imaging technology is the noise performance. Figure 4 compares the measured noise floor of HIT to that predicted by theory as a function of pixel rate. Both the hybridized detector array and the CMOS amplifier independent of the imaging array have been considered. The prototype CMOS amplifier has a noise floor rivaling that of state-of-the-art CCDs. Unfortunately the baseline noise floor of the amplifier was found to increase substantially after mating to the detector. The increase in noise floor seen in the hybridized device, as compared to the amplifier alone, is a result of a larger-than-expected parasitic capacitance contribution at the signal detection node. Understanding the magnitude and source of total parasitic capacitance allows the CMOS designer to optimize the amplifier design for that particular capacitance. The theoretical optimized line shown in Figure 4 represents the potential HIT performance that could be obtained by optimizing the CMOS amplifier design based on the actual parasitic capacitance.

Power usage is another major performance parameter that must be considered in any next-generation imaging technology. Figure 5 demonstrates the power advantage of the HIT approach by plotting power utilization versus data rate for three representative 1024×1024 imaging device approaches. Total power required for the analog data output can be predicted by considering the power usage of the amplifiers in conjunction with the drive power required for charge transfer through the array. In these calculations it has been assumed that each device has 16 output channels. For the case of the CCD, it is assumed that each on-chip source follower amp dissipates between 6 mW and 12 mW, depending upon data rate and load resistance. In the case of the HIT device, a 60 μW to 500 μW power usage is assumed for each preamplifier, with a 1 mW output buffer residing at each output channel. By comparing the CCD and HIT device power curves, the power benefit of the HIT approach is obvious. However, it is possible to further reduce the total power requirement of the imager. The third curve in Figure 5 represents power estimates for the next generation of HIT detector, the parallel hybrid ultra-low-noise detector (PHUD). For PHUD it is assumed that a 15 μW preamplifier mediates each column in the detector array and that a 1 mW buffer resides at each of 16 output buffer channels. It can be seen that parallel processing provides an opportunity for a sizable reduction in required power at high image rates.

(a)   (b)

Figure 3 *Typical images from the prototype 256 × 512 HIT. (a) The brightest portion of the image corresponds to a 50-electron signal level. No background subtraction or image enhancement has been performed on the image, which was taken at an operating temperature of - 40 C. (b) The right-hand side image is a 3000-electron peak image taken at 0 C, also without image correction.*

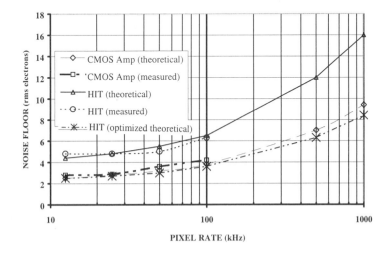

Figure 4 *Measured and predicted noise floors for the CMOS amplifier and the HIT detector*

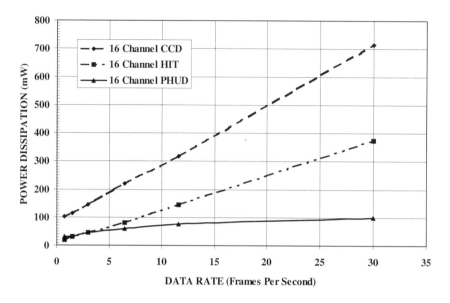

Figure 5 *Power versus frame rate for three 1024 × 1024 pixel detector architectures*

6 FUTURE DEVELOPMENT

The prototype HIT device has clearly demonstrated the feasibility of hybridizing CCD-type detectors to CMOS-based amplifiers. Having proven the concept, the next step in development is the incorporation of additional signal processing components, such as correlated double sampling, into the CMOS chip. In addition, ancillary CMOS timing, driver, and control chips will be required to complete the desired single package scientific-quality microcamera. These items are currently under design and development.

References

1.   B. Olson, T. Shaw, B. Pain, R. Paniccaci, B. Mansoorian, R. Nixon, E. Fossum, 'A
     Single Chip CMOS APS Digital Camera', *Proceedings of the 1997 IEEE Workshop
     on Charge-Coupled Devices and Advanced Image Sensors,* Bruges, Belgium, June
     5–7, 1997, pp. R24.1-R24.4.

INTEGRATING ELEMENTAL AND MOLECULAR IMAGING

Jon R. Schoonover, George J. Havrilla and Patrick J. Treado

Chemical Science and Technology Division, Los Alamos National Laboratory
Nuclear Materials Technology Division, Los Alamos National Laboratory
ChemIcon, Inc., Pittsburgh, PA

1 INTRODUCTION

The objective of our research effort has been to develop an array of spatially resolved techniques to bring together complementary elemental and molecular information. One goal is to integrate the data for a comprehensive picture addressing the composition of complex materials. The overall product of this effort is to provide a nondestructive, integrated characterization method for heterogeneous materials by producing elemental and molecular-level images that address issues related to sample composition. Such an approach represents a cost-effective method for obtaining data to both characterize radioactive, hazardous chemical, and mixed wastes as well as assess disposal and reduction methods by sampling before and after treatment.

The advantage of an array of micro-techniques is in minimizing sample size and handling, thereby reducing the cost and time required for characterization, and in reducing worker exposure to hazardous materials. The aim is to provide both elemental and molecular images of the material that will graphically illustrate compositional variance on a single sample. Elemental and molecular-level imaging approaches represent a new and emerging area of research with considerable promise for applicability to a wide variety of chemical problems.

The disposition and treatment of various waste forms contaminated with actinides are a major focus area within the U.S. Department of Energy. The study of interactions and structural characteristics of uranium, neptunium, plutonium, and americium species in different substrates is vital information in exploring stabilization, processing, and storage scenarios. The interactions of actinides with various waste forms and media, and the fate and transport of actinides, are areas of particular interest. In this paper, we focus on using microscopic imaging technology to study the behavior of plutonium in brine solution, characterizing an incinerator ash sample containing plutonium, and characterizing corrosion in a tantalum coupon exposed to a plutonium surrogate.

2 EXPERIMENTAL

2.1 Elemental Imaging

X-ray fluorescence (XRF) utilizes X-ray photons to eject inner core electrons from an element. The resulting fluorescence signal is indicative of a particular element. Spatial distribution, in our studies, is measured using micro-X-ray fluorescence (MXRF) instrumentation or scanning electron microscopy with energy dispersive detection.

The instrumentation used for MXRF studies is a Kevex Omicron spectrometer equipped with a 50W rhodium X-ray anode oriented at a 45-degree angle with respect to the sample stage. The detector is also oriented at a 45-degree angle. In MXRF mapping, a liquid-nitrogen-cooled, lithium-drifted silicon chip with an active area of 50 square mm is used to detect the fluorescence signal from the sample. Apertures are used to spatially restrict the X-ray beam with a minimum spot diameter of 30–50 micrometers. The mapping is accomplished by moving the sample stage and allowing the primary beam to scan the sample. The stage is moved at a user-selected speed. Scanning a selected number of frames and co-adding the intensity from each frame in a dynamic mode produces an elemental image.

Elemental analysis was also performed with a scanning electron microscope (SEM; RJ Lee Instruments) outfitted with secondary electron and backscatter electron detection capability. The elemental imaging was accomplished with the SEM using an energy dispersive spectrometer with point analysis and elemental mapping capability.

2.2 Molecular Imaging

Raman spectroscopy uses laser light scattering to measure the vibrational frequencies of molecules. Infrared spectroscopy involves the absorption of infrared radiation to provide vibrational information. These measurements provide a spectroscopic "fingerprint" that reveals the identity and condition (phase, oxidation state, and molecular structure and speciation information) of molecules that constitute a material. Vibrational imaging experiments based on these spectroscopic techniques utilize either direct imaging or mapping. Direct imaging is accomplished by illuminating a large field of view of the sample and imaging the scattered (Raman) or transmitted (IR) light onto an imaging two-dimensional array detector (e.g. a CCD camera; InSb or MCT focal plane array) positioned at the focal plane of the optical system. The mapping experiment involves measuring a spectrum at each point of a grid for well-defined spatial locations of a sample. An image can then be reconstructed from the spectrum at each location. In some instances, direct imaging and mapping are combined into hybrid methods, as in line scanning imaging Raman spectrometers.

We have utilized a number of different approaches to produce Raman images. The basic experimental arrangement for Raman microscopy and point-by-point or line-scan imaging utilizes laser excitation of 488.0 or 514.5 nm, supplied by a Spectra Physics 2025 Ar^+ laser or 752 nm from a Ti: Sapphire laser pumped by the 2025 Ar^+ laser. This laser excitation is coupled to an infinity-corrected optical microscope (Ziess Axiovert 135) using a holographic SuperNotch Plus rejection filter (Kaiser Optical Systems, Inc.). In the line scanning experiment, a mirror is operated in an oscillating mode to produce the excitation line. The microscope objective (Ziess LD-EPIPL 20x/0.4 or Ziess PLAN-NEO 63x/0.95) is used to both focus the laser light and collect the Raman scattering. The collected light is passed back through the holographic notch filter and dispersed by a HoloSpec f/2.2 monochromator (Kaiser Optical Systems, Inc.) equipped with the appropriate holographic Raman grating and a liquid-nitrogen-cooled Photometrics PN12 CCD.

IR microscopy and imaging (point-by-point mapping) in our laboratory uses a Spectra-Tech Research IRplan microscope, a Nicolet 20SXB FTIR bench, and a programmable Cell Robotics microscope stage. Synchronization of stage movement, data collection, and certain data processing steps are accomplished through the Nicolet MacrosPro software interface with samples examined in the transmission mode.

Direct Raman imaging experiments are conducted using the Falcon Raman imaging microscope system (ChemIcon, Inc.). The Falcon system uses a diode pumped $Nd:YVO_4$ solid state laser source doubled to operate at 532 nm (Spectra Physics, Millennia II) coupled with a multimode fiber optic relay to an infinity-corrected optical microscope (Olympus, BX60) via an illuminator assembly (ChemIcon, Raman Illuminator). The illuminator defocuses the laser source and excites the entire sample field of view through a 50X objective (Olympus, 0.80 N.A.). The Raman scattering is collected with the same objective and is transmitted back through the illuminator, which houses holographic notch rejection filters to remove the Rayleigh scattering. The Raman signal is filtered with a 9-cm^{-1} bandpass liquid crystal tunable filter (LCTF) constructed using the Evans Split-Element geometry (CRI, VariSpec). Raman images are collected using a TE cooled (-40 ^{o}C) slow-scan charge-coupled device (CCD) detector (Princeton Instruments, TE/CCD-512TKB) having 512×512 (20-mm-square) pixels.

2.3 Sample Preparation

One of the main considerations in integrating elemental and molecular imaging approaches is identifying a sampling matrix that is transparent (with minimal interference) to X-ray, infrared, and visible radiation. This matrix would enable the entire suite of techniques to interrogate the same specimen in a safe manner and facilitate correlation of multi-spectral data with visual features. Many of the systems we are interested in studying contain or have been in contact with actinides. Encapsulation or immobilization of the material, with minimal sample handling, is a necessity.

In the studies to date, we have compromised by attempting to immobilize the sample while changing sampling cells for the different imaging approaches. For the study of precipitated materials that have been in contact with actinides, the particle for examination was smeared onto a filter paper. The samples were then contained between two sheets of Mylar in an XRF sampling cell for elemental imaging. The sample was then removed from the XRF cell and sealed (using epoxy) between a microscope slide and cover slip for Raman measurements.

A second sample type we have studied is a few particles of a plutonium-contaminated incinerator ash. Because of the mobility of this sample type, it was smeared on a section of doubled-sided tape that was attached to a section of filter paper. This ash sample was then contained between two Mylar film layers for X-ray analysis; the specimen was then placed on a microscope slide and sealed under a glass cover slip for Raman analysis. For IR analyses, particles were typically removed from the sample and compressed between two 2-mm BaF_2 windows under mild pressure generated by a compression cell.

The third sample studied was scale material from a tantalum corrosion study. This sample was examined with no sample preparation. The sample was studied using a combination of SEM with energy dispersive detection and Raman imaging.

2.4 Data Acquisition

One approach we have taken is to utilize XRF maps or elemental images as an initial screening technique to identify regions of interest within a sample. In this approach, low-resolution elemental images over a large area serve to identify (1) correlations between different elements, and (2) the areas of interest within the sample. For many of our

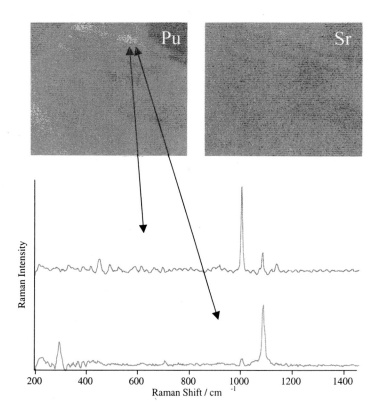

Figure 1 *Elemental images demonstrating the correlation between Pu and Sr along with Raman spectra in the region of high Pu concentration*

applications, this is an area of high plutonium or actinide concentration. Once these areas have been identified, high-resolution XRF imaging, along with Raman and IR microscopy and imaging, serve as probes to gain further insight into the composition of the material and the correlation between molecular species.

The spatial resolution of the MXRF method is limited to tens of micrometers. When higher spatial resolution is needed, SEM studies were utilized. The coupling of SEM and Raman imaging represents an approach with essentially diffraction limited spatial resolution.

3 APPLICATIONS

3.1 Highly Heterogeneous Precipitated Material

The study of chemical interactions and structural characteristics of actinide (uranium, neptunium, plutonium, and americium) species in different substrates is an area of significant interest in the context of exploring stabilization, processing, and storage

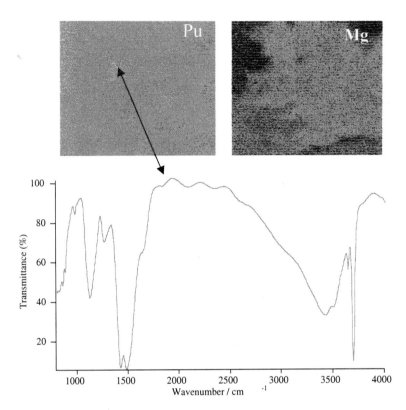

Figure 2 *Elemental images of Pu and Mg and IR spectrum in the area of high Pu concentration*

scenarios. Interactions of actinides with various media and their mobility in the media are areas of research where elemental and molecular imaging can play an important role. The Actinide Source-Term Waste Test Program (STTP) at Los Alamos is a project designed to study the behavior of actinides in brine.

In this project, test containers with actinide contaminated waste are exposed to brine solutions. These test containers are periodically tested for soluble actinides and precipitated material. The containers are kept at constant temperature, rotated weekly for 15 min, allowed to settle for 2–3 days, and sampled at selected intervals. Typically, 50 mL of brine solution is extracted from the test container and filtered in series through 5-μm, 1-μm, and < 20-nm filters.

The filter samples represent a highly heterogeneous sample type where combining elemental and molecular data can assist in characterizing the material. Elemental imaging of the filters is performed to correlate elemental composition within these precipitates, and to gain further information on the molecular species present; Raman and infrared measurements were then made on spatially distinct regions of the sample.[1]

Figure 1 is an example of one precipitate studied by this approach. Figure 1 shows the elemental images for the elements Pu and Sr. The analytical data (not shown) suggested that high concentrations of Pu correlated with increased levels of Sr and S, while Ca and Cl were major constituents. The images demonstrate a clear correlation between Pu and Sr. The two Raman spectra (200–1450 cm^{-1}) in Figure 1 are from 1 to 5-µm particles in the area of the precipitate sample that demonstrated high Pu levels. The Raman data for this particular precipitate indicates the presence of $CaSO_4 \cdot 2H_2O$, $SrSO_4$ (Figure 1, top spectrum) and $CaCO_3$ (Figure 1, bottom spectrum). Interestingly, the Raman data coupled the elemental images demonstrate the presence of $SrSO_4$ in the area of high Pu concentration leading to the interpretation that $SrSO_4$ can co-precipitate the Pu in this test container.

Figure 2 shows the utility of using IR coupled with elemental images to assess the precipitate from a different test container. The elemental images demonstrate the distribution of Pu and a high level of Mg, even though it is not a component of the brine solution. The IR data suggests that $Mg(OH)_2$ is the primary component of the precipitate.

3.2 Waste Material

An incinerator ash sample is another example of a Pu contaminated material where understanding the comprehensive composition can lead to a better understanding of the chemistry. The primary interest is to identify the Pu species present and their relationship to other elements or molecular components in the sample. The sample is incinerator ash of plutonium-contaminated waste originating from the Rocky Flats Environmental Technology Site, and is a fine gray-brown powder with black particles.[2]

Elemental maps of Pu concentration for a few particles of the ash sample demonstrate the distribution of Pu in this sample. The Raman and IR spectra were measured from areas of high Pu concentration and provide information of the Pu species and its relation to other inorganic complexes. The most intense feature in the Raman microscopy spectrum is at 475 cm^{-1}. Actinide oxides are known to crystallize at high temperature with a fluorite (CaF_2) structure and space group $Fm3m(O^5_h)$ that possesses one Raman active phonon of T_{2g} symmetry (478 cm^{-1} for PuO_2) at k = 0. Therefore, the 475-cm^{-1} Raman band can be assigned to this T_{2g} mode, while other oxide species (SiO_2, TiO_2, and Al_2O_3) also contribute to the Raman spectrum.

In Figure 3, a Raman image from the 475-cm^{-1} Raman band is correlated with an area demonstrating high Pu concentration from the MXRF image. This data provides a definitive assignment for the PuO_2 as the predominant Pu species in the ash sample. A direct Raman spectrum collected through the LCTF from one of the micrometer-sized particles is also shown. Comparison of the Raman and IR microscopy data of particles of high PuO_2 concentration further demonstrates that SiO_2 and PuO_2 are closely associated in this sample with a heterogeneous distribution of additional metal oxide species. As with the precipitated material, the elemental images serve to direct the analysis to the area of interest in the sample. The molecular microscopy and imaging then supply further information on these spatially distinct areas.

3.3 Corrosion

Tantalum is an excellent metal as a container for plutonium solutions. Tantalum is known to corrode under certain conditions, although the overall rate is slow. When comparing different corrosion conditions, it has been discovered that a scale material is formed during reflux of tantalum with nitric acid, hydrofluoric acid, and ceric ammonium nitrate as a surrogate for plutonium solution waste. In this particular experiment, a

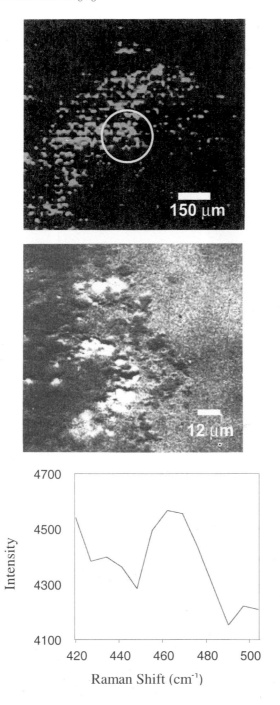

Figure 3 *Elemental image of Pu distribution with Raman chemical image of circled area and LCTF Raman spectrum of PuO₂*

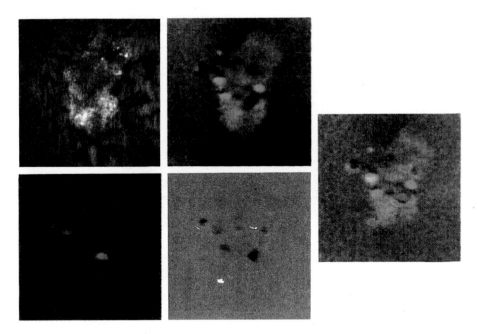

Figure 4 *A bright field image of the corrosion site with cosine correlation analysis images at 1308 (red), 661 (blue), and 965 cm^{-1} (green)*

tantalum coupon was refluxed for 100 h. in 300 mL of nitric acid, 1 mL of HF, and 32 g of $(NH_4)_2Ce(NO_3)_6$. Understanding the composition of the scale material can lead to a further understanding of the corrosion process.[3]

For the scale material, elemental mapping was performed using SEM/EDS and detected elements including O, Na, Ta, Cl, K, Ca, Ti, and Fe. The dominant elements in the scale were Ce, O, and F; these elemental images demonstrated co-location of Ce and F. Raman chemical imaging of the scale material provides high contrast molecular images of the corrosion product. With the raw data, Raman images were produce at 1308, 661, and 965 cm^{-1}, which corresponded to the most intense spectral features in the Raman spectrum from the corrosion product. Figure 4 shows images using these Raman shifts and processed using cosine correlation analysis (CCA). CCA is a multivariate technique that can reveal subtle spectral differences not obvious in the raw data.[4] These data, coupled with principal component analysis, have suggested four principal components in the scale. One component has been identified as CeF_3. The opposite side of the scale has also demonstrated the presence of Ta_2O_5. Studies are continuing and software strategies are being explored to gain further insight into this scale material.

4 CONCLUSIONS

Combining and integrating elemental and molecular imaging represents a new and powerful development in scientific optical imaging. In this paper, we have utilized this approach to interrogate highly heterogeneous material to begin to understand the underlying chemistry relevant to the storage and disposition of actinide contaminated material. This approach and this area of imaging will continue to expand and improve, allowing further insights into the composition of complex materials.

References

1. J. R. Schoonover, F. Weesner, G. J. Havrilla, M. Sparrow, and P. J. Treado, *Appl. Spectrosc.*, 1998, **52**, 1505.
2. J. R. Schoonover and G. J. Havrilla, *Appl. Spectrosc.,* 1999, **53**, 257.
3. C. T. Zugates, A. S. Bangalore, and P. J. Treado, 'Raman/SEM Chemical Imaging of Heterogeneous Inorganic Materials', ChemIcon Report, 1998.
4. H. R. Morris, J. F. Turner II, B. Munro, R. A. Ryntz, and P. J. Treado, *Langmuir*, 1999, **15**, 2961.

CCDS FOR THE INSTRUMENTATION OF THE TELESCOPIO NAZIONALE GALILEO

R. Cosentino, G. Bonanno, P. Bruno, S. Scuderi
Osservatorio Astrofisico di Catania
Viale Andrea Doria, 6
I-95125 Catania (Italy)

C. Bonoli, F. Bortoletto, M. D'Alessandro, D. Fantinel
Osservatorio Astronomico di Padova
Vicolo dell'Osservatorio, 5
I-35131 Padova (Italy)

1 ABSTRACT

Most of the scientific instrumentation, as well as the tracking systems and the Shack-Hartmann wavefront analyzers at the Italian National Telescope "Galileo", use charge-coupled devices (CCDs) as detectors. The characterization of detectors is of fundamental importance for their correct utilization in scientific instrumentation. We report on the measurement of the electro-optical characteristics of different CCDs that will be used in the scientific instrumentation at the Italian National Telescope. In particular, we will show and compare the quantum efficiency, the charge-transfer efficiency, the dark current, the read-out noise, the uniformity, and the linearity of two sets of CCDs manufactured by EEV and LORAL. Finally, we will show the preliminary tests performed at the telescope with the optical imager, which has a mosaic of two EEV chips.

2 INTRODUCTION

It is no secret that CCDs are the dominant detectors in instrumentation for optical astronomy. The Italian National Telescope is no exception to this rule. Three out of four of its instruments, as well as the cameras for tracking and adaptive optics, use CCDs as detectors.

The measurement of the electro-optical characteristics of these detectors is the necessary preliminary step to allow for proper selection of the detector and then to ascertain its optimal use at the telescope. We have characterized two sets of CCDs, manufactured by EEV and LORAL, selected for the instrumentation of the Italian National Telescope.

In Section 3 we describe the Italian National Telescope and its scientific instrumentation using CCDs. In Section 4 we outline the main characteristics of the CCD controller. Section 5 is devoted to a description of the facilities used for calibration of the CCDs. Section 6 deals with the results of measurements of the CCD characteristics. Finally, in Section 7, the results of tests performed at the telescope are shown.

3 THE ITALIAN NATIONAL TELESCOPE GALILEO

The Telescopio Nazionale Galileo (TNG) is the national facility of the Italian astronomical community and is located at Roque de los Muchachos (alt. 2400 m.) in La

Palma (Canary Islands, Spain). TNG is an altazimuth telescope with a Ritchey-Chretien optical configuration and a flat tertiary mirror feeding two opposite Nasmyth foci.

The diameter of the primary mirror is 3.58 m with a focal length of 38.5 m (F/11). The scale is 5.36 arcsec/mm and the vignetting-free field of view is 25 arcminutes in diameter. The telescope will host an optical and an infrared camera at Nasmyth focus A and two spectrographs at Nasmyth focus B.

3.1 TNG Scientific Instrumentation Using CCDs

As previously stated, CCDs are widely used at TNG for guiding, tracking, adaptive optics and, of course, for the scientific instruments. However, we measured only the characteristics of the CCDs used in scientific instrumentation. These instruments comprise the optical imager, the low-resolution spectrograph, and the high-resolution spectrograph.

3.1.1 Optical Imager Galileo. The Optical Imager Galileo (OIG) is the CCD camera for direct imaging at optical wavelengths (3200–11000 Å) for Galileo. It is placed at the Nasmyth adapter interface A and is usually illuminated by light coming directly from the TNG tertiary mirror to the f/11 focus. No other optical elements are in front of the CCD, apart from those on the filter wheels (filters, polarizers, and atmospheric dispersion correctors) and the dewar window. The OIG is designed to host a variety of CCD chips or mosaics covering a field of view up to 10 arcmin.

3.1.2 Low Resolution Spectrograph. The low-resolution spectrograph (LRS) is a focal reducer instrument installed at the spectrographic Nasmyth focus (Nasmyth adapter interface B) of the TNG. The LRS camera is opened to F/3.2. The scale is 0.276 arcsec/pixel with a field of view of 9.4 square arcmin. Currently, the observing modes available are direct imaging, long slit spectroscopy, and multi-object spectroscopy. The camera is equipped with a 2k × 2k LORAL CCD.

3.1.3 Spettrografo ad Alta Risoluzione per Galileo. The "Spettrografo ad Alta Risoluzione per Galileo" (SARG) is the white pupil cross-dispersed echelle spectrograph under construction for the TNG (Gratton et al 1998). SARG is a high efficiency spectrograph designed for a spectral range between 3700 and 9000 Å and with a resolution ranging from R=19,000 up to R=144,000. SARG uses an R4 echelle grating in Quasi-Littrow mode. The beam size is 100 mm, giving an RS product of RS=46,000 at the center of the order. Single object and long slit (up to 30 arcsec) observing modes are possible. A dioptric camera images the cross-dispersed spectrum onto a mosaic of two 4k × 2k EEV CCDs.

4 THE CCD CONTROLLER

The CCD controller that runs all CCDs at the TNG is described in detail in Bonanno et al., 1995, and Bortoletto et al., 1996. Here, we will give a brief description of its main characteristics.

4.1 The Control System Of The CCD Cameras

The architecture of the CCD's readout system is shown in Figure 1. The first block is the CCD controller, which is located close to the cryostat. In the CCD controller there is a bus (the CCD controller bus), into which are plugged a sequencer board and at least one analog board.

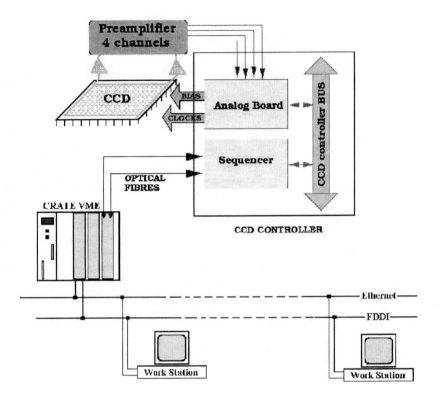

Figure 1 *CCD readout system*

The sequencer board generates the clock sequences. Each analog board produces eight programmable bias voltages and sixteen clock drivers with independently programmable upper and lower levels and is also able to read and process four channels independently.

The controller is connected to a VME crate through optical fibers. The VME contains a shared memory for image storing, a transputer to send commands to the controller and an Ethernet connection to the TNG workstation system.

4.2 Waveform Editor

To allow different readout modes, the CCD controller generates table-based sequences. These tables are easily generated by using a Windows program. This program, called *waveform editor*, allows generation of the *tables* through a graphical interface and "mouse actions". Figure 2 shows an example of a waveform editor session.

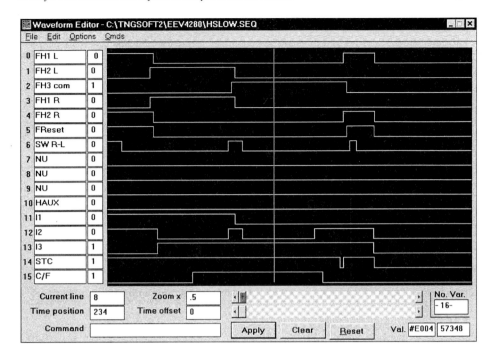

Figure 2 *Waveform editor*

5 CALIBRATION FACILITIES

The main calibration of CCDs used for the TNG instrumentation is performed at the Catania Astrophysical Observatory. The whole range of CCD electro-optical parameters can be measured using the facilities described in the following paragraph. We have developed a small laboratory at the telescope itself to tune up the CCDs and their operating electronics and to check the temporal behavior of their performances.

5.1 The Calibration Facility of the Catania Astrophysical Observatory

The Catania Astrophysical Observatory calibration facility allows a full electro-optical characterization of CCD detectors. The main component of the facility is the instrumental apparatus to measure quantum efficiency (QE) in the wavelength range 1300–11000 Å (see Figure 3). The radiation emitted by two light sources (deuterium and xenon lamp) passes through a series of diaphragms and filters and is then focused onto the entrance slit of a monochromator. The dispersed radiation beam is divided by a beam splitter and focused on a reference detector and on the CCD. A detailed description of this apparatus can be found in Bonanno et al., 1996.

For uniformity and linearity measurements, a 20-in integrating sphere is optically connected to the QE measurements system through a quartz singlet. The useful wavelength interval range in this case is 2000–11000 Å. Finally, system gain and CTE measurements are performed using a Fe^{55} X-ray source (Figure 4).

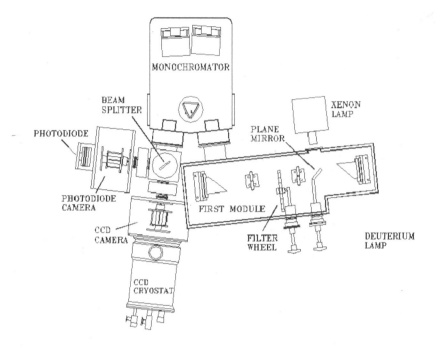

Figure 3 *Scheme of the instrumental apparatus*

5.2 The Calibration Facility at the Telescope

The calibration facility at the TNG detectors laboratory (see Figure 5) consists of a xenon lamp, two filter wheels with a series of interference filters and a series of neutral filters, a 20-in integrating sphere, and a reference photodiode.

Figure 4 Fe^{55} *X-ray camera*

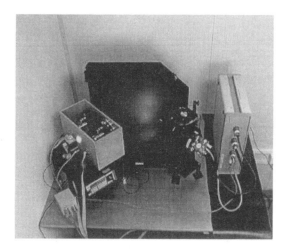

Figure 5 *Calibration facilities at the TNG*

6 CCD CALIBRATION

The CCDs used in TNG instrumentation are manufactured by EEV and LORAL. All of the chips are thinned back-illuminated with enhanced UV response. The LORAL chips are thinned at Steward Observatory, University of Arizona, and their UV response is enhanced using a technique developed there called chemisorption (Lesser and Venkatraman, 1998). The EEV chips are ion implanted. Table 1 summarizes the characteristics of the four CCDs (two from EEV and two from LORAL) for which the electro-optical performances have been measured. The two CCDs from EEV and one from LORAL are being used at the OIG, while the other LORAL CCD will be used at the LRS. For each CCD, we have measured system gain, read-out noise, linearity, uniformity, quantum efficiency, dark current and charge-transfer efficiency.

6.1 System Gain

The calculation of system gain, K (e⁻/DN), was performed using the X-ray stimulation technique. We used a Fe^{55} X-ray source that emits mainly 5.9 keV photons.

Table 1 *Manufacturer characteristics of TNG CCD*

Manufacturer	**LORAL**	**EEV 4280**
Chip Type	Thinned Back illuminated	Thinned Back illuminated
Pixel size (μm)	15	13.5
Area (pixel)	2048 × 2048	2048 × 4096
MPP	Yes	No
Working temperature	-110 C	-130 C
UV treatment	Chemisorption	Ion implantation
AR Coating	Yes	Yes
Grade	2 (both)	2 & 3

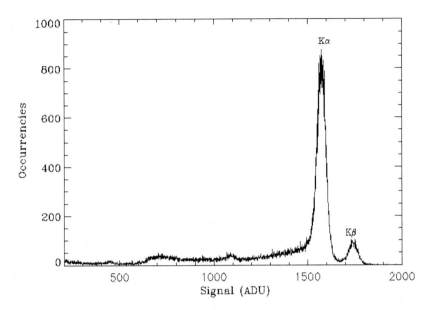

Figure 6 *Single pixel X-ray events—EEV 4280*

The X-rays produce an ideal point source of charge of known energy. On average, 3.65 eV of energy is needed to produce a single e-h pair, yielding an average of 1620 e⁻ produced in silicon per impinging X-ray photon. If the X-ray event occupies a single pixel, it is referred to as a single pixel event. Events that occupy more than a single pixel are referred to as split and/or partial events. Figure 6 shows the histogram of the single pixel X-ray events for one of the EEV CCDs (OIG). Both Kα (5.9 keV) and Kβ (6.5 keV) as well as Kα and Kβ escape peaks are clearly visible.

6.2 Read-out Noise

Read-out noise was measured as the RMS of the signal in the overscan region. The read-out noise was shown to be 12 e⁻ and 10 e⁻ for the LORAL and the EEV chips, respectively.

6.3 Linearity

Linearity was measured using a standard method of illuminating the CCD at different signal levels with a uniform source of radiation. The deviation from linearity measured for the two sets of CCDs are 0.15% for LORAL chips and 0.5% for EEV chips.

6.4 Uniformity

CCD homogeneity was measured by illumination with a uniform source of radiation at different wavelengths. The deviation from homogeneity is then given by the following expression:

Deviation from Homogeneity = RMS (whole area) / mean (whole area) %   (1)

Table 2 *Deviation from uniformity*

λ (Å)	LORAL	EEV 4280
4000	3.2 %	3.9 %
5500	3.4 %	3.4 %
7000	3.3 %	5.7 %
9000	3.7 %	5.2 %

Table 2 summarizes the deviation from uniformity for a LORAL and an EEV chip (OIG), while Figures 7 and 8 show two examples of flat field obtained at 9000 Å for the same chips.

6.5 Quantum Efficiency

CCD quantum efficiency (QE) was measured in the wavelength interval 2000-10500 Å in incremental steps of 500 Å. The CCD is illuminated using monochromatic light obtained through a xenon lamp, a series of filters and a monochromator. The signal integrated on the CCD is then compared to the response of a calibrated photodiode.

The major contribution to the absolute error on the QE is given by the calibration error of the photodiode, ranging from 5% to 7%, depending on the wavelength. Another source of error is found in the accuracy of the system gain measurement. Finally, one must take into account the fact that the image of the monochromator slit on the CCD is usually small compared to the dimensions of the whole sensitive area, making the non-homogeneity of the CCD another source of inaccuracy.

Figure 9 shows the QE of the two LORAL CCDs and of one of the EEV 4280 CCDs. It is interesting to note that the QE of the LORAL CCDs peaks at different wavelengths

Figure 7 *Flat field Loral CCD—900 nm*

Figure 8 *Flat field EEV 4280 CCD—900nm*

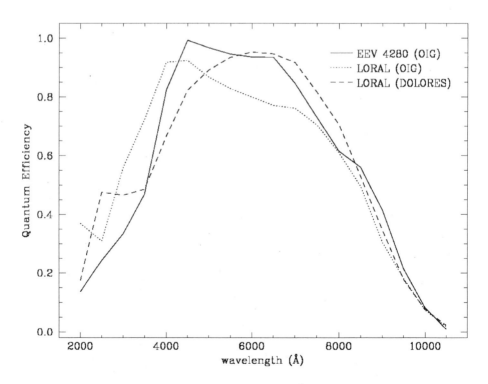

Figure 9 *QE measured in the wavelength 2000–10500 Å*

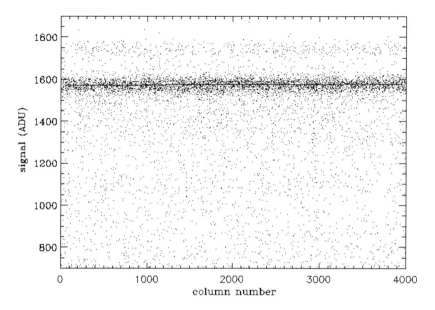

Figure 10 *X-ray spectrum*

(6000 Å for the LRS CCD and 4000 Å for the OIG CCD) due to differing thickness of anti-reflection (Hf$_2$O) coating (700 Å for the LRS CCD and 500 Å for the OIG CCD).

6.6 Dark Current

Dark current is measured by taking a short dark exposure and a long dark exposure. The signal level in the two exposures was then compared. The frame was divided in 10 × 10-pixel subframes, and the dark current was computed as the average of the values measured in each of the subframes. In the case of the EEV 4280, the average value of the dark current turned out to be 6 e⁻/pix/hour.

6.7 Charge Transfer Efficiency (CTE)

Charge-transfer efficiency (CTE) was measured using the X-ray stimulation method. In particular, the parallel CTE was obtained by exposing the CCD to the Fe[55] source for a given amount of time. After integration, the columns were stacked together and the signal plotted versus the number of pixel transfers. The CTE is then given by the following:

$$CTE = 1 - (\text{Charge loss})/1620 \text{ e}^-/DN. \tag{2}$$

One example of the results obtained is shown in Figure 10 for a LORAL (OIG) chip. The CTE of the LORALs was shown to be 0.999982, while the EEVs exhibited a CTE of 0.99999.

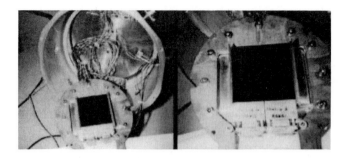

Figure 11 *EEV mosaic 4k × 4k mounted into the dewar*

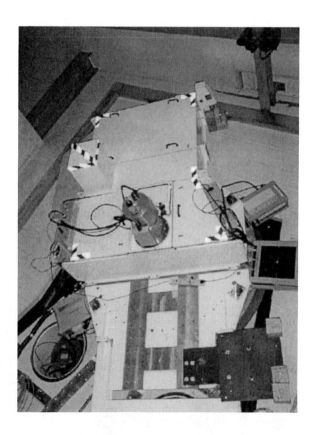

Figure 12 *OIG mounted at the Nasmyth focus*

Figure 13 *M15—V Johnson filter*

Figure 14 *NCG 2903—V Johnson filter*

7 TEST AT THE TNG

Tests at the telescope have been performed using the optical imager. We have chosen two EEV 4280s, placing them one close to the other, to form a 4k × 4k mosaic. Figure 11 shows the CCD mosaic, the cold finger, and the OIG dewar. The OIG dewar is placed at the Nasmyth A focus. Figure 12 shows the dewar mounted at the rotator adapter. At the time of the meeting, only preliminary tests had been done, basically "pretty" pictures of famous objects, of which two examples are shown in Figures 13 and 14.

The first is a picture of M15, taken in V Johnson filter, with an exposure time of 30 s. The second is the center region of NGC 2903, taken in V Johnson filter, with an exposure time of 60 s.

Acknowledgments

This work is based on observations made with the TNG telescope operated on the island of La Palma by the Centro Galileo Galilei in the Spanish Obervatorio del Roque de los Muchachos of the Instituto de Astrofisica de Canarias.

References

1. G. Bonanno, P. Bruno, R. Cosentino, F. Bortoletto, M. D'Alessandro, D. Fantinel, A. Balestra, and P. Marcucci, *IAU 167 Proceedings*, 1995, 319.

2. G. Bonanno, G., Bruno, P., Calì, A., Cosentino, R., Di Benedetto, R., Puleo, M., and Scuderi, S., *SPIE Proceedings*, 1996, **2808**, 242.

3. F. Bortoletto, C. Bonoli, M. D'Alessandro, D. Fantinel, G. Farisato, G. Bonanno, P. Bruno, R. Cosentino, G. Bregoli, and M. Comari, *SPIE Proceedings*, 1996, **2654**, 248.

4. R. Gratton, A. Cavazza, R. U. Claudi, M. Rebeschini, G. Bonanno, A. Calì, R. Cosentino, and A. Scuderi, *SPIE Proceedings*, 1998, **3355**, 769.

5. M. P. Lesser, I. Venkatraman, *SPIE Proceedings*, 1998, **3355**, 446.

SPECTRAL IMAGING WITH A PRISM-BASED SPECTROMETER

Jeremy M. Lerner

LightForm, Inc.
Belle Mead NJ 08502 USA

1 INTRODUCTION

Imaging spectroscopy has its origins in the remote Earth monitoring community. It is well known that mineral deposits, ground water absorption, pollution, the health of agricultural and forest products, and karyotyping of human chromosomes can be identified from spectral information.[1,2,3,4] Experience has shown that greater spectral detail enables more detailed information to be generated and with it the enhanced ability to discriminate between objects that otherwise appear similar.

There are two major methods of acquiring spatially resolved wavelength data: multiple wavelength bandpass filters or wavelength dispersive systems such as diffraction grating or prism[5]. Using either acquisition method, maps are built up by overlaying spectral data onto images acquired using conventional means, such as a photograph or CCD camera. For images that require up to 20 wavelength data points per spectral object, the process is referred to as *multispectral imaging* and for images needing up to 200 wavelength data points (or more), the process is known as *hyperspectral imaging*. In all applications where hyper- or multispectral imaging provides a solution, the field of view of the sample is heterogeneous. Just as the surface of the Earth is a complex mixture of objects, so too are the cells and tissues found in the biological and medical sciences.

2 SPECTRUM AND IMAGE CUBES

2.1 Spectrum Cubes

When filters are used to generate hyperspectral images, an entire field-of-view (FOV) is captured simultaneously and viewed through a succession of filters incrementally, as shown in Figure 1a. This technique is intensely processor and memory hungry with file sizes typically ranging from 10 to 200 Mb, even with relatively simple acquisitions. If the sample requires significant deconvolution analysis, then file sizes grow even larger. It is not unusual, therefore, to hear that a Cray computer was used to analyze the wealth of data. This is not to say that desktop computers cannot be used in simple cases, but the data processing of a hyperspectral image is almost always off-line and time consuming.

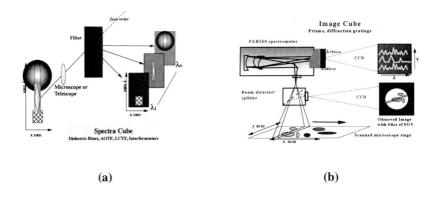

(a)   **(b)**

Figure 1 *(a) Creating a spectra cube, and (b) creating an image cube*

Filters used in this type of imaging spectroscopy range from simple dielectric filters on a rotating carousel to acousto-optic tunable filters (AOTF), liquid crystal tunable filters (LCTF), and interferometers.[6,7]

2.2 Image Cubes

Image cubes are acquired by a wavelength dispersive system that incorporates a diffraction grating or prism. These instruments typically require an entrance aperture, usually a slit, which is imaged onto the focal plane of a spectrograph at each and every wavelength simultaneously. Such an instrument, essentially free of aberrations, can determine the origin of each and every source of photons on the slit. Therefore, an object imaged on the slit will be recorded as a function of its entire spectrum and its location in the sample. Figure 1b illustrates a prism-based imaging spectrometer interfaced with a microscope. The sample under analysis is on a microscope slide, which is translated so that objects are projected onto the entrance slit and "pushed" across it. This is often referred to as a "push-broom" acquisition method to generate the image cube, illustrated in Figure 2b.

The prime advantage of this method is that all the wavelength data needed to identify an object or objects, even if the spectra are highly convoluted, are acquired simultaneously and are immediately available for processing. This technique is ideal for kinetic studies, samples that exhibit movement, studies of time-based changes in molecular characteristics, or any condition that benefits from real-time spectral analysis and reporting. Because a single push-broom acquisition is always modest compared to the formation of a spectrum cube, a PC handles even complex samples with 750 wavelength data points.

Figures 2a and 2b illustrate how "image" and "spectrum" cubes present spatial and spectral information.

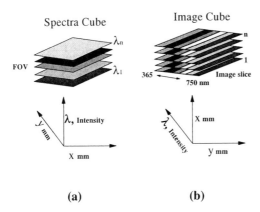

Figure 2 *(a) Representation of a spectra cube, and (b) representation of an image cube*

3 EQUIPMENT

Traditionally, the easiest way to acquire a particular field of view was with a filter-based instrument. This was due mostly to the poor optical quality and transmission efficiency of wavelength dispersive systems such as those based on a diffraction grating. The use of newer, highly specialized prism spectrometers has enabled the design of imaging spectrometers with almost perfect imaging and transmission efficiency in the visible averaging 90%.

The system described here consists of a PARISS® Imaging Spectrometer and accompanying software (LightForm, Inc., Belle Mead, NJ), (illustrated in Figure 1b); a Nikon Eclipse E400 epi-fluorescence microscope with dual beam splitter (Nikon Inc., Melville NY), with a Chroma Technology Corp (Brattleboro, VT) 31002-filter-cube; a computer-controlled translation stage on the microscope (Prior Scientific Inc., Rockland, MA); and two EDC 1000L CCD cameras (Electrim Corp, Princeton NJ). One camera captures spectral data from the imaging spectrometer and the other acquires an "observed" image corresponding to the field of view seen through the microscope eyepiece.

4 SOFTWARE CHALLENGES

4.1 Overview

When a spectrum cube is generated, each pixel on the CCD detector chip acquires a wavelength data point for each filter acquisition. Therefore, each pixel accumulates a spectrum with as many data points as there are acquisitions. To give some idea of the file size that a spectrum cube generates, assume that an object can be identified with a spectral resolution (bandpass) of 5 nm [at the Raleigh criterion, the full width at half its maximum (FWHM) = 5-nm]. Therefore, 1 data point per nm provides the data points needed to resolve a spectral feature with a 5-nm FWHM. A 10-Mb file accommodates 55 data points (over a spectral region of 55 nm), when acquired with a low-resolution 180 K pixel CCD camera, before the first data processing computation can begin. The software program must then interpret and associate each spectrum with an object in the field of view (FOV). Considering that a heterogeneous sample may demand spectral acquisitions between 360–750 nm to characterize all objects, a spectrum cube presents an enormous data processing and memory burden.

By contrast, image cube acquisitions are much more flexible, because all the wavelength information needed to make a decision is contained in each individual acquisition. It is not necessary to generate a huge file before the first data processing computation can begin. An imaging spectrometer interfaced to a 180,000-pixel, 8-bit, CCD chip produces a 180-Kb file with each acquisition. A file of this size is easily managed with a PC with close to real-time data processing. With the aid of color-coding techniques, entire spectra can be compressed into a single word of memory.

For many applications in medicine, as well as the life sciences, it is not necessary to create complete images. Often a push-broom approach is preferred because it enables point-and-shoot acquisitions, commonly employed when a target can be easily identified or when the target is moving. A spectrum cube generated by a filter device is not effective for either a moving target or for real-time delivery of information concerning a particular specimen.

4.2 Software Functions

The hyperspectral imaging system described in this paper integrates many dependent and independent functions:

- Spectral and observed image acquisition, each with its own CCD camera.
- Exposure time, gain, and bias of each of two cameras.
- Background correction as well as standard and proprietary math functions.
- The use of false-color enhanced images to more easily observe high and low S/N ratio spectra or images.
- Creates spectral Regions of Interest (ROI).
- Records each of 242 spectra acquired simultaneously in terms of intensity and location at the sample.
- Codes each spectrum and enables sophisticated correlation routines to compare known spectral fingerprints with acquired spectra.
- Operates a translation stage to enable a sample to be automatically "pushed" across the entrance slit of the spectrometer.
- Wavelength calibration and correlation to pixel position.
- Data reporting.

5 IMAGE ACQUISITIONS

The core objective of hyperspectral imaging is to acquire spectra and correlate a spectral fingerprint with a particular object or condition. In the examples below, we illustrate how a complex observed FOV is characterized both spectrally and spatially.

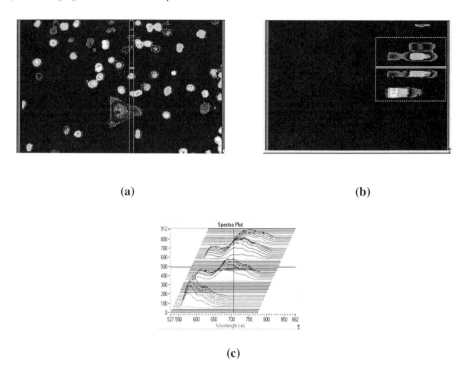

(a)   (b)

(c)

Figure 3 *(a) Observed image of a field of fluorescing pollen grains; the rectangular dotted slit in the center field shows a bright horizontal strip corresponding to that in Fig. 3b; (b) Spectral image of the pollen grains contained by the slit in 3a; the x-axis is wavelength and the Y-axis location in the slit; and (c) Spectral detail corresponding to Figure3b*

5.1 Acquisition Characteristics

- Pixels in wavelength direction (x-axis):   752
- Pixels in spatial direction (non-interlaced in y-axis):   242
- Number of data points per spectrum: 752
- Number of spectra per acquisition: 242
- Wavelength coverage (simultaneous):   385–750 nm

5.2 Defining a Region of Interest (ROI)

The spectrum CCD, the wavelength detector for the spectrometer, has a chip with 242 rows of pixels, each of which acquires a spectrum. There are 750 pixels to a row, so every spectrum can be defined by up to 750 data points. Positions on the y-axis correlate with the location of light sources on the spectrometer entrance slit. Many fluorescence acquisitions do not require full wavelength coverage. In addition, not all objects in the FOV may be of interest. The program enables the user to select the wavelength region of choice, as well as a key spatial region on the entrance slit. If the user chooses to acquire less than 242 spectra with each acquisition, rows may be combined or " binned". For example, 242 rows can be presented as 121 "strips" each two rows wide over the wavelength range from 500–600 nm. A "strip" is one or more combined rows of pixels.

5.3 Image Acquisitions

Figure 3a shows a false-color enhanced observed image with a dotted slit passing vertically through the image. This slit can be thought of as a projection of the entrance slit of the spectrometer acting like a window through which a portion of the sample passes to the imaging spectrometer. Figure 3b shows the false-color enhanced spectral image presented by the CCD detector of the spectrometer. These images are shown on the program's standard desktop. The x-axis corresponds to wavelength and the y-axis location on the entrance slit. The spectral Region of Interest (ROI) is indicated by the large rectangle in figure 5b. Figure 3c shows the spectral plots defined by the ROI.

5.4 Classification of Spectra

To gain maximum flexibility, we use two classification methods: one supervised and the other unsupervised or "self-classifying". When an object presents very obvious spectral characteristics, the user has the option to select that region of the spectrum and paste it into a spectral library. Spectra acquired in this way can be color coded so that any recurrence in subsequent acquisitions will immediately show in the selected color(s).

Spectral libraries can accommodate as many spectra as the user wishes, although it becomes a challenge to visually differentiate between numerous colors.

5.4.1 Supervised Classification. In this mode, the user selects a limited ROI, either by dragging or digitally entering the coordinates of an area that represents a key spectral feature. The spectrum defined by this area is then entered into the library and the process repeated until the field has been defined. Each spectrum is referred to as a "class".

5.4.2 Unsupervised Classification. In this mode, the user selects an ROI that accommodates all objects under analysis. The computer then automatically classifies all spectra within the ROI. All similar spectra are combined into a defined "class". The user controls the sensitivity by limiting the number of classes the program is permitted to find. This process sets the sensitivity threshold of the analysis. A histogram slider enables the background, baseline, or saturated pixels, to be subtracted so that noise below a certain level is not assigned to a class of spectrum.

The maximum number of spectral classes is the number of strips in the ROI; so if 242 strips are selected, it is possible to generate a very large number of spectral classes. However, the program limits the number of classes to those that meet a minimum correlation coefficient (MCC). Typically, the larger the number of classes, the better the correlation. If there are 100 strips and 100 classes, then the MCC will equal 1.0. In practice, the user is able to subtract the background and low level noise to enhance the signal to noise (S/N). This typically enables a high MCC with a limited number of spectral classes.

Experience has shown that the number of classes can best be selected by deciding on the MCC in advance and adjusting the number of classes until the goal is met. Putting this function under user control is an invaluable tool to enable a researcher to rapidly characterize a sample without unnecessarily large spectral data libraries. Following unsupervised classification, the spectra can be transferred to a permanent "master spectrum" library file. Figure 4 shows the master spectrum file generated when we used

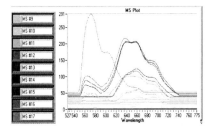

Figure 4 *Master spectra library generated by "unsupervised" classification from Figure 3b*

unsupervised classification to define the sample shown in Figure 2a. Out of a total of 120 spectra, the program determined that 17 distinct spectral classes were present. These color-coded spectra contributed the colors in Figure 3c.

A library generated from unsupervised classification is almost always large; therefore it can be a challenge to visually discriminate between shades of color-coded spectra. The user can judiciously edit the library and remove spectra considered unnecessary.

5.5 Correlation of Spectral and Observed Image Data

It is not until the acquired spectra are correlated with an observed image that the process is complete. Because of inevitable small mismatches, either in path length through beam splitters or ancillary optics, it is virtually impossible to guarantee perfect correlation between a spectrum and its precise location on the observed image produced by the second CCD. This may not pose a problem for most applications but can be a crucial consideration. We addressed this problem in a very simple way. Each acquisition is stored in three formats concurrently:

- As a *classified image,* produced by the imaging spectrometer and presented as a color-coded spectrum image.
- As a *processed image,* generated by acquisition by the imaging spectrometer and presented as a false color image as a function of intensity.
- As a *combination image*, which is an overlay of the classified and processed images.

5.5.1 The Classified Image. Each strip, or combined groups of rows, contains a spectrum. The program attempts to correlate this spectrum to a library of color-coded spectra. If the correlation criteria are met then the color code of the associated spectrum is returned. If there are 242 strips then there will be 242 color elements presented in a vertical column. Figure 5a shows an example in which the black areas show no correspondence with a spectrum in the library. (For clarity we enlarged the line width from 1 pixel to 10 pixels). This column is also referred to as a "chromagram". As the projected image of the sample is translated across the slit, each acquisition concatenates the next chromagram to the last. The image continues to build-up at the operator's discretion.

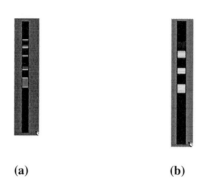

(a)   **(b)**

Figure 5 *(a) Spectral "chromagram" and (b) intensity map*

Example 1: Given that the ROI is defined by: $x = 0$ to 750 and $y = 0$ to 242, and the number of strips = 242, then each of the 242 rows is individually summed. Each row presents a false color determined by the magnitude of the sum. Therefore, a column is produced, 242 pixels high, composed of 242 color elements one pixel wide by one pixel high. The colors in the chromagram are determined by the spectrum contained in each of the 242 rows when correlated with the spectra in the master library.

- If an acquired spectrum does not correlate with a spectrum in the library, to equal or exceed the MCC, then that element in the column is black.
- If the MCC is met or exceeded, then the program displays the color associated with that spectrum in the library.

Example 2: Given that the ROI is defined by: $x = 400$ to 700 and $y = 0$ to 240, and the number of strips = 20. Each of the 20 strips correlates with a spectrum in the library, each color element (including black for a spectrum that does not correlate) is, therefore, 12 pixels high and one pixel wide in a chromagram 240 pixels high.

5.5.2, The Processed and Observed Images. The Processed Image and the Observed image are functionally the same. The observed image is acquired through the second CCD camera and presents a full image of the FOV, as shown in Figure 3a. The processed image is built up acquisition by acquisition during the push-broom process by summing each row of pixels within the ROI. The resultant values are false-color coded as a function of their intensity. Each false color element is then recorded in a column to exactly match the column produced in the classified image above, as shown in Figure 5b. (For clarity we enlarged the line width from 1 pixel to 10 pixels).

Example 1: Given that the ROI is defined by $x = 0$ to 750 and $y = 0$ to 242, and the number of strips = 242, then each of the 242 rows is summed. Each row presents a false color determined by the magnitude of the sum. Therefore, a column is produced 242 pixels high composed of 242 color elements one pixel wide by one pixel high.

Example 2: If the ROI is defined by $x = 400$ to 700, and $y = 0$ to 240, and the number of strips = 20, then a column of 242 pixels is formed based on the sum of each row over the range of pixels from 400 to 700, each element is one pixel high and one pixel wide. The column is 242 pixels high regardless of the number or strips.

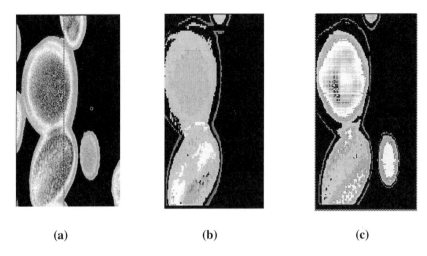

(a)   (b)   (c)

Figure 6 *(a) Observed image showing a false color enhanced pair of large pollen grains under fluorescence excitation, magnification = 40X; all colors are a function of fluorescence intensity; (b) processed image of Fig. 6a; all colors are a function of intensity; and (c) classified image presenting a spectral topographical map of 40 classes of spectra. Note: the changes in color in this figure are NOT related to intensity. Each color represents a unique spectrum*

5.5.3 The Combination Image. As the pushbroom process continues, the classified and processed images are created simultaneously, acquisition by acquisition, until the area of the sample under investigation is completed. The operator has the choice of five false color algorithms to maximize contrast between the processed image and the classified image when the two are overlaid. The combination of processed and classified images ensures that there is 100% perfect pixel-to-pixel matching when comparing spectra versus real image.

6 HYPERSPECTRAL IMAGING

Pollen provides good examples of objects with multiple fluorescent signatures. An observed image was recorded of a limited field of pollen grains through the second CCD camera and false color enhanced as shown in Figure 7a. The hyperspectral images were formed by concatenating chromagrams created in classified and processed image modes. The ROI was selected to cover the spectral range from 545 and 780 nm, corresponding to 260 pixel data points. In the spatial y-axis, we used the full height of the image with 242 rows of pixels, which was divided into 121 strips, 2 pixels in height and 260 pixels in width.

A spectral acquisition was made through the centerline of the two pollen grains and the spectra defined by unsupervised classification. We made the decision that the acquired spectra must correlate with the spectra in the sample with an MCC of 98%. To achieve this goal, we required 40 spectral classes. These were then entered into a master

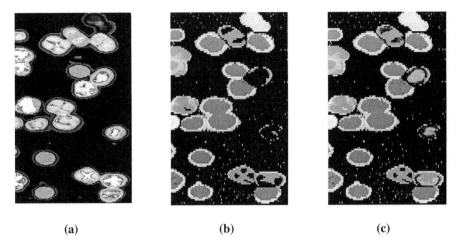

(a)   (b)   (c)

Figure 7 *(a) Processed image of a population of pollen grains, false color enhanced, taken through the imaging spectrometer; all false colors are a function of intensity; (b) classified image presenting the spectral topography of the same pollen grains; just as isobars can be color coded to map areas of equal pressure, the colors in this figure show areas with the same spectral characteristics; and (c) combination image forming a perfect overlay of (a) and (b)*

library for use during subsequent acquisitions. Each acquisition generates 240 spectra corresponding to the number of rows of pixels in the ROI. Each pair of rows was combined to form one row resulting in 120 strips within the ROI and a 120-point chromagram. With a 40X objective, this corresponds to a spectrum, each 1 μm, along the pollen grains in the sample. The slide was scanned in 1-μm increments to form Figures 6b and 6c and required a total of 127 acquisitions resulting in 127 concatenated chromagrams.

Figure 6a, shows false color enhanced observed image of a pair of large pollen grains, magnified by 40X, with some smaller pollen grains in the vicinity. Figure 6b shows the processed image, acquired as described in Section 5, through the imaging spectrometer and false color enhanced as a function of intensity. As expected, the morphology of the images presented in Figures 6a and b are different due to path differences between the CCD on the imaging spectrometer and the second CCD camera.

Figure 6c shows the spectral topography of the two pollen grains. The oval object at the bottom right is present in the observed and processed images, but absent in the classified image because its spectral characteristics did not correspond with any spectrum in the master library.

To obtain relative wavelength intensity data or to study a particular area of the sample in greater detail, the program enables the user to click on a particular target area on the Classified Image and the computer sends the stage to that location.

Figures 7a and 7b show a population of pollen grains magnified by 10X. Classified and processed chromagrams were acquired every 4 μm at the sample. Changes in color in Fig. 7a correspond to changes in spectral characteristics. Again, about 40 spectral classes were required for a MCC of 98%. Figures 7a and 7b required a total of 152 acquisitions resulting in 152 concatenated chromagrams.

If these spectral images had been acquired by a filter method, each would conservatively occupy about 40 Mb compared to approximately 150 Kb for the push-broom method presented in this paper.

7 DISCUSSION

We have been able to demonstrate that hyperspectral images can be obtained from bio-fluorescent materials using an elegant system interfaced with a standard microscope. We were able to correlate and map newly acquired spectra with up to 40 stored spectral signatures each defined by 250 data points. The hyperspectral images enable pure 1:1 pixel correlation with the equivalent video images.

The ability to perform sophisticated spectral topography offers researchers the opportunity to perform complex analysis on applications that include co-localization studies, detecting and mapping multiple fluorescent tags, FRET, FISH, live cell assays, epitope identification and mapping, and identification of inter/intracellular features.

Acknowledgements

Many thanks to Lew Drake for his software expertise and to Lee Stein for his help with the project and the preparation of the manuscript. We would also like to thank Stan Schwartz at Nikon Inc., Melville, NY, for the loan of the Eclipse 400 microscope and for his support.

References

1. E. Schrock, S. Du Manoir, T. Veldman, B. Schoell, J. Weinbert, M. A. Ferguson-Smith, Y. Ning, I. Bar-Am, D. Soenkesen, Y. Garini, and T. Reid, *Science*, 1996, **273**, 494.
2. J. M. Lerner, T, Lu, and S. Vari, *Proc. SPIE*, 1998, **3261**, 224.
3. J. M. Lerner and L. Stein, *Laser Focus World*, Oct. 1998.
4. D. F. Heath, Krueger, H. A. Roeder, and B. R. Henderson, *Opt. Eng.*, 1975, **14**, 323.
5. D. W. Warren and J. A. Hackwell, *Proc. SPIE*, 1989, **1055**, 314.
6. E. S.Wachman, W. H.Niu, and D. L. Farkas, *Appl. Opt.*, 1996, **35**, 5520.
7. Hoyt, *Biophotonics International*, July/August, 1996, 49.

COMBINING LINEAR AND NEURAL PROCESSING IN ANALYTIC INSTRUMENTS–OR WHEN TO SWITCH ON YOUR NEURONS

Nigel M Allinson and Boris Pokric
Department of Electrical Engineering and Electronics, UMIST

Edmund T Bergström and David M Goodall
Department of Chemistry, University of York

1 INTRODUCTION

The development drivers for enhancing scientific analytic instruments are to lower the detection limit, to increase throughput, and to provide more automated operation. The first two goals are clearly related—increasing sensitivity implies increasing acquisition times—but throughput can also be increased by deriving on-line the physical parameters of interest to the end user. As instrument designers, we can exploit improvements in sensor technology, optimize optical and mechanical design, and employ the latest advances in signal processing.

This paper concentrates on the last aspect through a case study of a novel instrument for *capillary electrophoresis* (CE), an increasingly important tool in separation science. The optimum recovery of low-level signal profiles embedded in noisy data is a generic problem and, hopefully, our technique may provide some pointers for applications in other fields. The approach is to combine the strengths of conventional linear filtering and adaptive non-linear function approximation using *radial basis function* (RBF) neural networks, together with the best possible estimate of system noise. There are, for our application, severe processing time limitations imposed by the short integration time of the detector array. Implementing the entire signal processing chain on a dedicated digital signal processing (DSP) sub-system makes it feasible to achieve optimum profile extraction within this integration time; so as far as the user is concerned, the analytic parameters of interest (e.g., diffusion coefficients, concentrations) are available in real time. All identified CE peaks are statistically valid to a user-specified confidence (e.g., 99% significance level).

2 CAPILLARY ELECTROPHORESIS FUNDAMENTALS

Over the last decade, CE has proved to be a versatile and rapid technique for the identification of water-soluble analytes, ranging from simple inorganic ions to macromolecular assemblies such as viruses.[1,2] As a technique, it offers high resolution, high analysis speed, simple preparation methodologies, small sample volumes, and potential automation. CE separation is based on the different electrophoretic mobility of

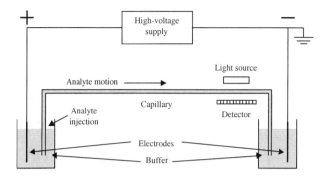

Figure 1 *Basic structure of a capillary electrophoresis instrument*

particles due to their varying charge and size ratios under the influence of an external electric field. A schematic of a CE instrument is given in Figure 1. A thin fused-silica capillary is filled with an aqueous buffer. The capillary ends are inserted into reservoirs filled with the same buffer and containing electrodes connected to a high voltage supply. The analyte is injected at one end of the capillary. Under the influence of the electric field, any charged particles will drift toward the oppositely charged electrode at rates depending on their charge-to-size ratios. The buffer will also move along the capillary because of electro-osmosis, the net result usually being that all particles travel in the same direction. At the non-injection end of the capillary, the temporally separated analytes are detected by measuring the relative changes in the absorbance or fluorescence due to the analytes as they pass across a detection window. A more detailed diagram of the CE system, when used in the *laser induced fluorescence* (LIF) mode, is provided in Figure 2.

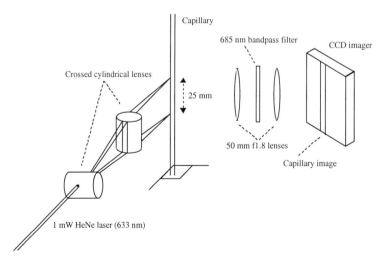

Figure 2 *Optical and detector arrangement for a CCD CE instrument in the laser induced fluorescence mode*

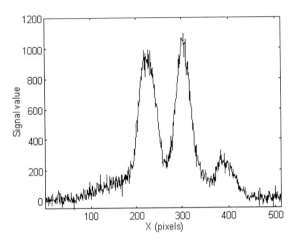

Figure 3 *Typical LIF profile of di-sulfonated aluminium phthalocyanin (AlPcS₂)*

The advantages of using an area charge-coupled device (CCD) imager include high sensitivity (coated with lumogen for good UV response), low noise, and high read-out rate. The same imager can be used either for multi-spectral absorbance measurements, employing a deuterium lamp light source and inserting a diffraction grating into the optical path, or for operation as a single-wavelength, one-dimensional imager, employing an optical filter and binning vertical columns for increased sensitivity and dynamic range. A typical LIF profile is shown in Figure 3. Further details of the instrument are already published.[3,4]

3 RADIAL BASIS FUNCTION NEURAL NETWORKS

An RBF network consists of a set of adaptive basis functions; the responses of each to an input signal are linearly combined through a second layer of learnt weighting factors to form the final output response.[5,6] The essential architecture of an RBF network is illustrated in Figure 4. The non-linear mapping between input and output is performed by basis functions that are normally Gaussian in form. The output of the *i*th basis function unit is

$$\phi_i(\mathbf{x}) = \exp\left(-\frac{\|\mathbf{x} - \mathbf{m}_i\|^2}{2\sigma_i^2}\right),\tag{1}$$

where $\mathbf{x}$ is the *n*-dimensional input signal vector, $\mathbf{m}_i$ is the position of the *i*th unit and σ_i is its width.

· The response of the *k*th output unit is given by

$$y_k = \sum_{j=1}^{J} w_{jk}\phi_j + w_{0k},\tag{2}$$

where w_{jk} is the weighting factor for the jth basis unit on the kth output unit and w_{0k} is the bias weight for this output unit.

The RBF network differs in several respects from the more widely known artificial neural network, the *multilayer perceptron* (MLP), not least in that the basis function units in the RBF network form a localized representation of the input space. This makes them potentially more suitable for our goal of optimally estimating the CE profiles within a strict processing time constraint. In general, RBF networks are easier to train than MLPs, as it is possible to modify the positions and spreads of the basis functions separately from training the weights in the final layer.

4 WHY NEURAL NETWORKS ARE NEEDED

If the CE profile peaks are broad, that is, composed of only low spatial frequencies, then it is simply necessary to employ linear low-pass filtering to accurately recover the profile (as the frequency spectra of the profile peak and the system noise are adequately separable). If the profile peaks are sharp, however, then their spectral bandwidth will overlap the system noise spectra sufficiently to make profile recovery impossible using merely linear processing. The adaptive non-linear filtering possible using the RBF network will permit sufficiently accurate profile recovery in this more demanding situation. The obvious question to ask is "Why not use the additional power of the RBF network for all situations?" If there were not strict limitations on the available processing time, then this would be possible. As explained later, the overall number of basis function units in the RBF network is built up one at a time. A broad peak would potentially require a large number of such units and the processing time could become

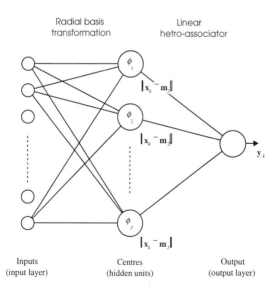

Figure 4 *Architecture of a radial basis function neural network*

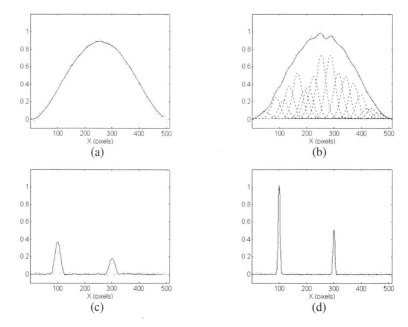

Figure 5 *Linear filtering and RBF network approximation of profiles. Top row is a broad (narrow bandwidth) profile and the bottom row is a sharp (high bandwidth) profile. Left hand column (solid lines) shows effect of low pass digital filtering. Right hand column (solid lines) shows effects of trained RBF network approximation.*

excessive[†]. An indication of the relative performance of linear filtering and non-linear RBF approximation for synthetic broad and narrow peaks is illustrated in Figure 5.

5 OVERVIEW OF CE PROCESSING

The area CCD detector produces video streams: one-dimensional for LIF operation through column binning and two-dimensional multi-spectral data for UV absorbance operation. The total length of the detection window is 25 mm with 512 photoelements or pixels, each approximately 50 μm wide. The main elements of the signal processing and parameter derivation chain are shown in Figure 6, and can be delineated into the following stages:

- Extraction of intensity changes across the detection window and best estimate of system noise;
- SNR enhancement using synchronous integration of successive CCD video streams (i.e., time delay and integration);
- Optimum profile approximation using linear or non-linear processing;
- CE peak identification; and
- Analyte velocity and diffusion coefficient calculation.

[†] It would be possible to make the individual widths of each basis adaptable, as well as their positions and amplitudes; but this additional freedom would markedly increase processing time.

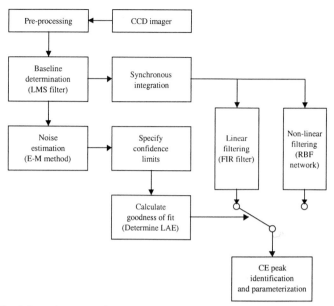

Figure 6 *Block flow diagram of major signal processing and parameter estimation stages*

6 BASELINE AND SYSTEM NOISE ESTIMATION

The raw video stream data must be corrected for non-uniformities across the detection window. Non-uniformity in the response of the individual CCD photoelements, uneven illumination, and variations in the "no analyte" transmission of the capillary are all factors. This correction can be achieved by monitoring the data when there are no CE peaks present. At the same time, system noise can be estimated. The "no analyte" or baseline signal could be calculated by averaging several video streams. The random system noise should tend to zero, leaving only the baseline signal. However, cross-correlating successive video streams after baseline removal does not result in a zero-mean cross-correlation signal, indicating that, among several possible effects, illumination and capillary transmission vary over time. An adaptive filter, with the error signal defined in terms of the cross-correlation product of successive video streams and trained using a modified least-mean square algorithm, is employed to estimate the current baseline correction (Figure 7). With this modified correction, the remaining signal for the "no analyte" condition represents the system's true zero-mean noise performance, as illustrated in Figure 8. Not including this adaptive filter and merely assuming that the baseline can be determined by simple averaging can result in profile approximation errors of up to 10%.

As the subsequent processing requires a good estimate of the noise characteristics, the probability density function (PDF) of the system noise is estimated using the well-known expectation-maximisation (EM) algorithm.[7] The EM method models the underlying PDF of a data sample as a mixture of simple Gaussian PDFs, where the nature of this mixture (number of Gaussians and their amplitudes and spreads) is estimated using an iterative maximum-likelihood algorithm. The estimated PDF of the

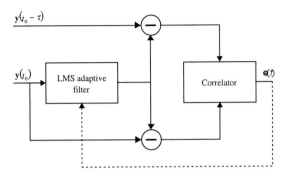

Figure 7 *Adaptive cross-correlation filter to remove residual baseline. The successive normalised video streams are* $\mathbf{y}(t_0 - \tau)$ *and* $\mathbf{y}(t_0)$

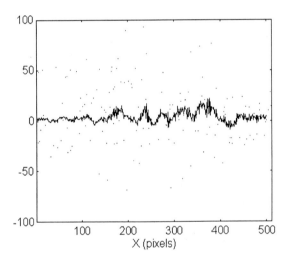

Figure 8 *Typical residual baseline (solid) and noise-only profile (dots)*

system noise is modelled in Figure 9 as three Gaussians. Attention to obtaining on-going, well-founded estimates of noise characteristics allows the setting of meaningful confidence limits for all subsequent processing and parameter values.

For the multi-spectral absorbance mode, it is necessary, in addition, to identify the optimum spectral band out of the twenty available in our instrumentation for further processing. This band is normally where the UV absorbance is the greatest.

7 SIGNAL ENHANCEMENT USING SYNCHRONOUS INTEGRATION

The CCD video streams can be considered as a set of contiguous snapshots where CE peaks are captured several times as they migrate across the detection window. As the rate at which these peaks move is known from the experimental conditions, it is possible to

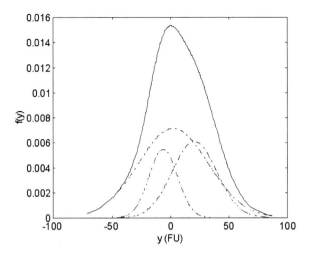

Figure 9 *Estimated system noise PSD derived using the E-M algorithm. The presence of three Gaussian functions to accurately represent the noise implies a number of independent noise sources*

accurately overlay successive video streams so that the CE peaks constructively add while the uncorrelated noise component is suppressed. This method is similar to the well established *time delay and integrate* (TDI) operation of imaging sensors in remote sensing systems where the known relative motion between the observation satellite and the earth is exploited to enhance the images. The normalized integrated signal, $\bar{y}_a(t)$, at the time interval t is given by the following:

$$\bar{y}_{a_i}(t_0) = \frac{y_i(t_0) + \bar{y}_{a_k}(t_0 - \tau)}{N_i(t_0)} \qquad \text{for } i = [0, K-1] \qquad (3)$$

where $k = i - s_i(t_0 - \tau)$

s_i is the shift parameter determined by the external experimental conditions,

K is the number of elements (pixels) in the image output,

τ is the camera frame time,

and $N_i(t)$ is the ith element of a K-element signal integration array $\mathbf{N}(t)$, which maintains a count of the summations for each element.

Each element's SNR is enhanced by a factor $\sqrt{N_i(t)}$. As there are more snapshots available of a peak at the non-injection end of the detection window, the noise suppression is greater at this end, as can be seen in the scatter of data points in Figure 10, with the injection end towards the right of the figure. This variable averaging across the detection windows means that many processing routines have to be modified by including a χ^2 correction term. Any form of signal averaging carries the potential risk that profile morphology could be blurred. In the prototype instrument, this is not so, as peaks approaching the theoretical minimum width are readily observable.

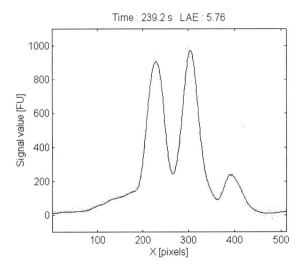

Figure 10 *CE profile after synchronous integration and low-pass filtering*

8 ADAPTIVE PROFILE APPROXIMATION

Initially, linear filtering, using the Parks–McClellan optimal equiripple design,[8] is attempted to estimate the underlying signal profiles. The FIR low-pass filter possesses a linear phase response within its passband ($f_{cut-off}$ = 13 pixels^{-1}), which ensures that the morphology of broad peaks is not corrupted. Furthermore, the common problem of filter artefacts due to finite filter length is suppressed by employing a boundary data matrix constructed from data points acquired at time instance t_0-τ. The filter response, $\mathbf{y}_a$, for each pixel position is given by the following:

$$\mathbf{y}_a = \left(\hat{\mathbf{Y}} + \hat{\mathbf{B}}\right)\mathbf{h}^{\mathrm{T}} \tag{4}$$

where $\mathbf{h}$ are the t filter coefficients $\mathbf{h} = \begin{bmatrix} h_1 & h_2 & \cdots & h_t \end{bmatrix}$,

   $\hat{\mathbf{Y}}$ is the profile data matrix acquired at time instance t_0

and $\hat{\mathbf{B}}$ is the boundary data matrix.

An example of a linear filtered profile is illustrated in Figure 10.

   The deviation of the filtered signal and the corresponding input signal is constantly monitored. If this deviation is less than the current system noise, then we can assume optimal profile extraction. The *approximation error*, ζ, is calculated for each sample point (pixel), as follows:

$$\zeta_j = \begin{cases} 1 & for \ f\left((y_{aj} - \hat{y}_j)\sqrt{N_j} \pm \sqrt{y_{aj}}\right) > f(\eta_r) \ and \ y_{aj} - \hat{y}_j > 0 \\ -1 & for \ f\left((y_{aj} - \hat{y}_j)\sqrt{N_j} \pm \sqrt{y_{aj}}\right) > f(\eta_l) \ and \ y_{aj} - \hat{y}_j \le 0 \\ 0 & for \ f\left((y_{aj} - \hat{y}_j)\sqrt{N_j} \pm \sqrt{y_{aj}}\right) \le f(\eta_r) \ and \\ & for \ f\left((y_{aj} - \hat{y}_j)\sqrt{N_j} \pm \sqrt{y_{aj}}\right) \le f(\eta_l) \end{cases} \tag{5}$$

Where y_{aj} are the approximating function points,

$\hat{y}_j$ are the corresponding profile points,

N_j is the signal integration index,

$f(y)$ is the noise PDF, and

η_l and η_r are the chosen lower and upper threshold boundaries of the current estimated noise PDF. Their values are calculated by setting a confidence limit λ, using the following relation:

$$\int_0^{\eta_r} f(y)dy = \int_{\eta_l}^0 f(y)dy = \frac{\lambda}{200} \quad where \quad 0\% \le \lambda \le 100\% \tag{6}$$

The $\pm\sqrt{y_{aj}}$ term represents the additional multiplicative shot or photon noise component, which is subtracted in the absorbance and added in the fluorescence mode. The local approximation error (LAE) is given by the following:

$$LAE = \sum_{j=i+1}^{k} \zeta_j^2 \quad \forall \text{sign}(\zeta_j) = \text{sign}(\zeta_{j-1}) \ for \ \max(k-i) \tag{7}$$

When the LAE is zero, then all individual profile values are estimated to within an error less than the system noise. This condition is very strict, and a small tolerance can be permitted so that impulsive noise artefacts do not invalidate the processing. A more robust and practical tolerance limit is defined in terms of a maximum number of successive profile points for which the LAE has the same sign. This number can be established by calculating the theoretical minimum peak width using available knowledge of the injection conditions, capillary length and (if known) diffusion coefficient estimates.

If the approximation error goes outside this limit, then the profile is approximated using the non-linear processing capabilities of the RBF network—that is, *you switch your neurons on.*

9 ON-LINE DESIGN OF THE RBF NETWORK

The elements of training a RBF neural network were presented earlier, but we need to ensure that its design (in terms of the number of basis units, their placement and parameters) is exemplary and that the computational effort is minimized. Potentially, basis units can be positioned at any pixel location—in our case, all 512 of them! Major speed improvements can be achieved by considering only those positions where a basis unit can meaningfully contribute to the final approximating function. These positions can be identified by taking into account the estimated system noise and the TDI integration index.

Optimization of a RBF network depends on minimizing a *cost function*, C, of the general form:

C = Sum of squared differences between individual data samples and the
             corresponding approximating function value
+ Sum of product of individual regularization parameters and second-layer
             weightings

The regularization parameters determine the trade-off between the smoothness of the approximation and its closeness of fit to the actual data. For LMS supervised training, the need is to determine the minimum of the cost function:

$$C = \sum_{i=1}^{p} \left(\hat{y}_i - f(\mathbf{x}_i) \right)^2 + \sum_{j=1}^{m} \lambda_j w_j^2 \tag{8}$$

where $f(\mathbf{x}_i) = \sum_{j=1}^{m} w_j h_j(\mathbf{x}_i)$

$\{w_j\}_{j=1}^{m}$ second layer weights,

$\{h_j\}_{j=1}^{m}$ basis functions,

$\{\lambda_j\}_{j=1}^{m}$ regularization parameters, and

$\{(x_i, \hat{y}_i)\}_{i=1}^{p}$ paired training set.

The minimization of C leads to a set of m simultaneous linear equations, for which the optimum set of weights can be expressed as follows:

$$\hat{\mathbf{w}} = \mathbf{A}^{-1}\mathbf{H}^T\hat{\mathbf{y}} \tag{9}$$

where **H** is the design matrix:

$$\mathbf{H} = \begin{bmatrix} \mathbf{h}_1 & \mathbf{h}_2 & \cdots & \mathbf{h}_m \end{bmatrix}$$

$$\begin{bmatrix} h_1(x_1) & h_2(x_1) & \cdots & h_m(x_1) \\ h_1(x_2) & h_2(x_2) & \cdots & h_m(x_2) \\ \vdots & \vdots & \ddots & \vdots \\ h_1(x_p) & h_2(x_p) & \cdots & h_m(x_p) \end{bmatrix}$$

and $\mathbf{A}^{-1}$ is the variance matrix, defined as

$$\mathbf{A}^{-1} = \left(\mathbf{H}^T\mathbf{H} + \Lambda\right)^{-1}$$

$$\text{where } \Lambda = \begin{bmatrix} \lambda_1 & 0 & \cdots & 0 \\ 0 & \lambda_2 & \cdots & 0 \\ \vdots & \vdots & \ddots & \vdots \\ 0 & 0 & \cdots & \lambda_m \end{bmatrix}$$

The approximated output, $\hat{y}_i$, for the ith component, is given by:

$$\hat{y}_i = f_i = f(\mathbf{x}_i) = \sum_{j=1}^{m} \hat{w}_i h_j(\mathbf{x}_i) = \overline{\mathbf{h}}_i^{\mathrm{T}} \hat{\mathbf{w}} \tag{10}$$

$$\text{where } \overline{\mathbf{h}}_i = \begin{bmatrix} h_1(\mathbf{x}_i) \\ h_2(\mathbf{x}_i) \\ \vdots \\ h_m(\mathbf{x}_i) \end{bmatrix}$$

The full approximated output, $\mathbf{f}$, for all p training pairs can be expressed as follows:

$$\mathbf{f} = \begin{bmatrix} f_1 \\ f_2 \\ \vdots \\ f_p \end{bmatrix} = \begin{bmatrix} \overline{\mathbf{h}}_1^{\mathrm{T}} \hat{\mathbf{w}} \\ \overline{\mathbf{h}}_2^{\mathrm{T}} \hat{\mathbf{w}} \\ \vdots \\ \overline{\mathbf{h}}_p^{\mathrm{T}} \hat{\mathbf{w}} \end{bmatrix} = \mathbf{H} \hat{\mathbf{w}} \tag{11}$$

Our strategy in building the RBF network employs a technique called *forward selection*, where basis units are added one at a time until some criterion is met. Forward selection is a non-linear algorithm that possesses several advantages, including no predetermined maximum number of units and low computational complexity. The first unit is that which most reduces the cost function; the second is that which most reduces the residual cost function, and so on. The basic algorithm can be speeded up by ensuring that each new basis unit is orthogonal to all previous ones. An additional improvement is to re-estimate the regularization parameters after each new unit is added. We stop adding basis units when the estimated prediction error is at a minimum (or at least asymptotic to a constant value). The prediction error estimates how well the current trained network will perform with future (i.e., as yet unknown) inputs.

To confirm the successful operation of the trained network, some form of validation procedure is required. A traditional approach in neural network applications is to split the input data into training and test sets and to use the test set to evaluate the trained network performance. This approach suffers in two respects: it requires an "excess" of data and may unwittingly introduce a bias in the splitting of data into the two sets. This simplistic cross-validation can be greatly improved by training the network on combinations of the input data. Therefore, for p data samples, p networks are trained using all combinations of $(p - 1)$ samples with the remaining sample used for testing. The final error is the average of the individual mean-square errors. Such a method is time consuming; a more effective approach is to employ the *generalized cross-validation* (GCV) criterion.[9] The GCV involves the adjustment of the average mean-square error over the entire training set. In building the RBF network, we monitor the GCV and the LAE, adding basis units until one of the following conditions is satisfied:

- GCV stops decreasing,
- LAE is satisfied or approaches a constant value, or
- Maximum number of units is reached (i.e. we run out of processing time).

If the GCV criterion is satisfied before the LAE one, then the network is ill conditioned, and the basis function spread needs to be adjusted and the building process repeated. The following rules are applied in adjusting this spread:

- if m approaches M (i.e. number of basis functions in the design matrix is large), the basis spread needs to be increased, or
- if m is small the basis spread needs to be decreased.

A typical CE peak approximated using seven Gaussian basis units in the RBF network is shown in Figure 11. The majority of the signal processing is implemented on an Analog Devices SHARC DSP board hosted by the instrument's PC. Typically, 10 ms is required to recover the profile if linear filtering is sufficient and 120 ms if a RBF network needs to be developed, trained and validated. There can be up to fifty basis functions for a 512-element data frame, with sufficient processing time available using the dedicated signal processing card to perform up to approximately fifteen learning iterations.

10 CALCULATING THE SEPARATION PARAMETERS

Once the profile has been approximated, the CE peaks are identified and their parameters

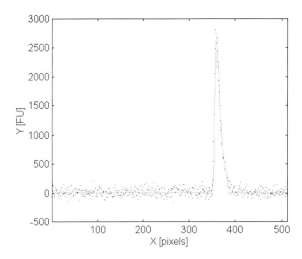

Figure 11 *Sharp profile approximation using a RBF network with seven Gaussian basis units with centres at pixel numbers 349, 356, 359, 361, 362, 366 and 376*

derived. Peak identification uses the first and second derivatives of the approximating function to locate local minima and maxima as well as points of maximum curvature. Hence, individual peaks are geometrically separated from the background and adjoining peaks. Each peak is parameterised in terms of its position, height and width. The number of identified peaks in successive snapshots may not be constant, because of noise

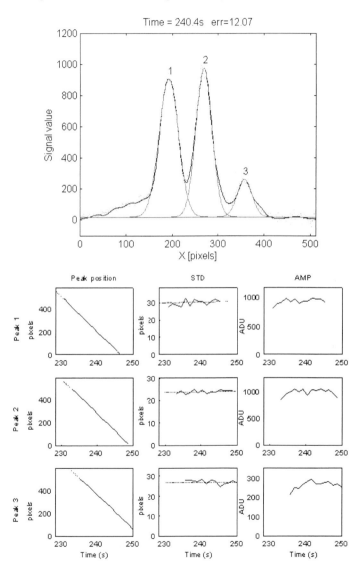

Figure 12 *Fully processed profile showing three identified peaks together with on-line tracing results that indicate each peak's position, standard deviation and amplitude. The dashed lines in the position and standard deviation graphs are linear regression lines that provide estimates of mobility and diffusion coefficient, respectively.*

artefacts and so forth, so it is necessary to track valid peaks. These peaks are identified using a velocity histogram method and specifying a maximum permissible amplitude change. In this way, any spurious impulsive artefacts can be eliminated. This peak identification is performed for each snapshot, allowing the differing mobilities, concentrations and diffusion coefficients (calculated from the peak widths) of each peak to be determined using a modified linear regression fit to take into account the varying confidence limits due to the different TDI integration indices across the window. Figure 12 shows three identified peaks and their estimated parameters.

11 CONCLUSIONS

Well-engineered signal processing based on current best practice and an intelligent application of different techniques can yield major improvements in this important class of analytic instrument. By incorporating such real-time processing within the instrument, more automatic operation can potentially be provided. For example, it is not possible to measure diffusion coefficients to any high degree of precision with a moving CE peak, as the spatial resolution needed to detect the slowly broadening peak would be impracticably fine. The conventional approach is to manually observe the migrating peaks and then stop the flow (by switching off the high voltage) and record the "stationary" peak over a long period. With real-time identification of peaks and calculation of diffusion coefficients, it is possible to pre-specify the required accuracy of measurement, detect a peak, stop the flow for the minimum time, and record the diffusion coefficient, all without operator intervention.

As stated in the introduction, our approach to profile extraction and peak identification could find applications in other areas. Hopefully we have demonstrated the complementary roles of linear and neural processing and the essential need to make the best estimate of a system's noise characteristics. By referring all decisions back to this noise, we can be confident that all identified signal peaks are statistically valid. A fuller account of the combined linear/non-linear adaptive signal processing is available.[10]

Acknowledgements

This work has been supported by the UK Engineering and Physical Science Research Council ROPA and Analytical Science grants, the latter in association with LSR AstroCam Plc.

References

1. P. Camilleri, 'Capillary Electrophoresis', CRC Press, Boca Raton, 1993.
2. R. Kuhn and S. Hoffsette, 'Capillary Electrophoresis: Principles and Practice', Springer-Verlag, New York, 1993.
3. D. M Goodall, E. T. Bergstrom, N. M. Allinson and B. Pokric, *Analytical Chemistry*, in press.
4. B. Pokric, N. M. Allinson, E. T. Bergstrom, D. M. Goodall, *J. Chromatography*, 1999, **A833**, 231.
5. · D. Lowe and A. R. Webb, *Network*, 1990, **1**, 299.

6. S. Haykin, 'Neural Networks: A Comprehensive Foundation', Prentice Hall, New Jersey, 1999.
7. S. V. Vaseghi, 'Advanced Signal Processing and Digital Noise Reduction', Wiley Teubner, Stuttgart, 1996.
8. E. C. Ifeachor and B. W. Jaris, 'Digital Signal Processing: A Practical Approach', Addison Wesley, Reading, 1993.
9. P. Craven and E. Wahba, *Numerical Mathematics*, 1979, **31**, 377.
10. B. Pokric, N. M. Allinson, E. T. Bergstrom and D. M. Goodall, *IEE Proc. on Vision, Image and Signal Processing*, in press.

ICIMACS: HOW WE GO FROM 0.3 TO 3 μ WITH 1 TO 40 AMPLIFIERS

Bruce Atwood, Jerry A. Mason, Daniel P. Pappalardo, Kevin Duemmel, Richard W. Pogge and Brian Hartung

Imaging Sciences Laboratory
Astronomy Department
The Ohio State University
Columbus, Ohio 43210-1106

The Imaging Sciences Laboratory (ISL) of the Astronomy Department of the Ohio State University has had an active instrumentation program for more than a decade. We are fortunate to have a group of ten full time permanent employees dedicated to the design and construction of instruments for ground-based astronomy in 0.3- to 2.5-micron wavelength range. Our instruments share a highly modular Instrument Control and Image Acquisition System, *ICIMACS*. The eight ICIMACS instruments operating on eleven different telescopes at seven observatories are listed in Table 1.

Each system consists of the following:

- The instrument, with its mechanisms for positioning the optical elements and detector(s);
- An instrument electronics (IE) package with mechanism drive electronics and a single board PC for control;
- One or more head electronics (HE) package(s) to control the detector(s), interface with the 120 MHz fiber link to the control room, control detector temperature, control the shutter in CCD systems, and provide serial communications interfaces to the instrument electronics and other telescope mounted devices, such as an etalon controller;
- A 120 MHz fiber-optic link for each HE between the telescope focus and the control room;
- One or more PC-based instrument computer(s) (IC) to house the ISL-designed sequencer that generates the digital pattern to control the detector, receive data from the HE, transfer data to the IC via direct memory access, and manage communications between the IC and the HE;
- A workstation computer (WC) (now a misnomer) to buffer the data and serve as a communications hub, and, in some cases, to interface to the telescope control system;
- A link to the local telescope control system, either from the WC or from the Sun;
- A Sun Sparc station or Linux computer with the user interface and data analysis programs.

Table 1 *Instrument and detectors currently operating under The Ohio State University ICIMACS Instrument Control and IMage ACquisition System (IC)*

OBSERVATORY	TELESCOPE	INSTRUMENT	DETECTOR(s)
Lowell Observatory	1.8-m	CCD Imager	2048 × 2048 SITe CCD
Kitt Peak National Observatory	4.0-m and 2.1-m	ONIS Near IR Imager/Spectrometer	512 × 1024 InSb Aladdin Array
MDM Observatory	2.3-m and 1.3-m	TIFKAM Near IR Imager/Spectrometer	512 × 1024 InSb Aladdin Array
MDM Observatory	2.3-m and 1.3-m	CCD Spectrograph	1200 × 800 CCD
Cerro Tololo Inter-American Observatory	4.0-m and 1.5-m	OSIRIS Near IR Imager/Spectrometer	1024 × 1024 HgCdTe HAWAII array
Cerro Tololo Inter-American Observatory	1.0-m YALO	Andicam Near IR/CCD Imager	1024 × 1024 HgCdTe HAWAII array *and* 2048 × 2048 LLICK3 CCD
South African Astronomical Observatory	1.0-m Elizabeth	Dandicam Near IR/CCD Imager	1024 × 1024 HgCdTe HAWAII array *and* 2048 × 2048 LLICK3 CCD
Wise Observatory	1.0-m	CCD Imager	2048 × 4096 SITe CCD
Michigan State University	various	CCD Imager	2048 × 2048 Thomson CCD

Figure 1 illustrates the layout of the instrument systems with a single HE and Andicam and Dandicam have two ICs, each connected to a single WC via separate SCSI busses, two fiber optic links and the functional equivalent of two HEs in a single package.

ICIMACS has successfully operated a wide range of detectors as shown in Table 1. Included are both CCD and infrared detectors having from two to thirty-two outputs. The HE uses a single ISL-designed printed circuit board for *all* the electronics connected to a detector. This printed circuit board can support the analog circuitry required to process the signal from up to 32 outputs. In addition, the HE can house a second, smaller, detector electronics board configurable for up to eight channels. Andicam and Dandicam operate in this two-detector mode.

All commands between the various nodes in the system are human readable and suitable for transmission over serial lines. A flexible message addressing and self-mapping scheme allows the system to be configured without concern for the specific identity of communication links. 9600-baud RS232 links are used throughout, with the exception of the Ethernet link between the WC and the Sun/Linux box, and, in some installations, Ethernet between the Sun and the telescope control system.

Data transfer between the IC and the WC is via a SCSI bus with controllers in each computer and two disks. The IC writes data on one disk while the WC is reading earlier

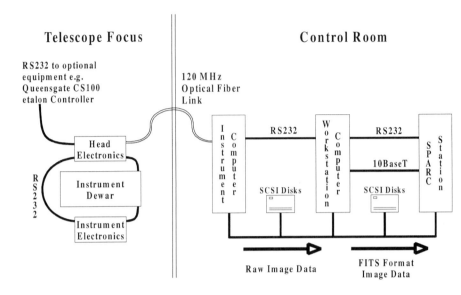

Figure 1 *ICIMACS architecture for single IC/HE systems. Andicam and Dandicam ha⋅ two ICs, two fiber optic links, and the functional equivalent of two HEs in a sing⋅ package*

data from the other disk. When the WC has read all available data, it requests that the I⋅ swap disks. When the IC is between frames it will begin writing data to the disk ju⋅ emptied by the WC and will advise the WC of the available data on the other disk. Data written to the disks in a very simple header + image format outside the operating syster⋅ This scheme allows transfers at nearly the SCSI hardware limit since the resulting di⋅ accesses are always to sequential sectors and tracks.

The same dual-controller two-disk SCSI bus is used to transfer data between the W⋅ and the Sun. A task on the Sun picks up data from these shared disks and writes into t⋅ Sun operating system. The net throughput of the system is high enough that $2k^2$ imag⋅ are available in the Sun operating system for analysis less than three seconds after detect⋅ readout. Our SCSI hardware is currently operating in the slowest and narrowest mod⋅ Upgrades with standard hardware could increase the speed of the transfers by as much as⋅ factor of 16.

Software on the IE, IC and WC is written in Power Basic and operates under DO⋅ Simple assembly language calls allow us to use a full 32 bit flat addressing model f⋅ transfer of large data blocks. The user interface on the Sun is written in C and provid⋅ complete status information, command interface and an extensive scripting capabilit⋅ IRAF and VISTA are available on the Sun (any other reduction package could I⋅ installed) and since data is left in FITS format in the Sun file system it is directly availab⋅ to any standard reduction package.

Our current main instrumentation project is the MODS twin dual-channel faint/mu⋅ object optical spectrographs for the 12-meter Large Binocular Telescope. Each of the fo⋅ MODS cameras will be equipped with two 4096 × 4096 pixel detectors.

The ISL welcomes the opportunity to provide other groups with more informati⋅ about our approach and designs.

ELECTRO-OPTICAL CHARACTERIZATIONS OF THE CID 17PPRA, CID 38SG, CID 38Q-A, AND CID 38Q-B

Q. S. Hanley,[†] J. B. True,[‡] M.B. Denton[*]

[†]Department of Molecular Biology, Max Planck Institute for Biophysical Chemistry, Am Faßberg 11, D-37070 Göttingen, Germany

[‡]Eastman Chemical Company, Texas Eastman Division, P.O. Box 7444, Longview, TX 75607-7444 USA

[*]Department of Chemistry, University of Arizona, Tucson, AZ 85721 USA

1 INTRODUCTION

The charge-injection device (CID) is a light-sensitive silicon detector belonging to the charge-transfer device (CTD) family of detectors. It consists of a continuous array of individual detector elements called pixels. Its unique capabilities distinguish it from the other subset of charge-transfer devices, the charge-coupled device (CCD).

1.1 History of CIDs

Gerry Michon and Hugh Burke invented the charge-injection device in 1973 while designing solid-state memory devices at General Electric.[1] Their original memory device had problems with photosensitivity, so they decided to pursue development of the CID as a photodetector.[2] After test devices of one and 1024 pixels 32 × 32, they produced a 100 × 100 device that was incorporated into a commercial camera system in 1973. In 1976, the first scientific use of a CID occurred at Kitt Peak Observatory.[3] Bonner Denton and his research group introduced the CID to analytical chemistry as a detector for emission spectroscopy[4,5]

In 1987, General Electric spun its CID development off to CID Technologies Inc., the current sole manufacturers of CIDs. They have developed a variety of detector formats and sizes, ranging from a 1 × 2 mm single pixel device CID75,[6] to a 768 × 512 CID22.[7] Larger formats with random-access CIDs have found use in such diverse applications as high-speed image tracking,[8] atomic emission spectroscopy,[9] and radiation hard video cameras.[7] The CID's flexibility is rooted in the simplicity of its device architecture.

The CID is based on metal oxide semiconductor (MOS) technology. The basic MOS structure consists of a metal conductor (gate) separated from a semiconductor by an oxide-insulating layer. The conductor in charge-transfer devices is not always metal but is more often polycrystalline silicon (polysilicon), highly p-doped for good conduction. Polysilicon is chosen for its greater optical transparency. The insulator is a silicon dioxide layer approximately 20 nm thick. CIDs designated Safeguard (e.g. 38SG) also have a silicon nitride passivation layer deposited on top of the oxide layer to ensure that no holes exist in the oxide layer. Devices without the silicon nitride layer are all-oxide devices, designated by a Q (e.g. 38Q). The semiconductor layer, more commonly referred to as the epitaxial layer, or epi, is an n-doped layer of silicon 15–20 μm thick. Unlike CCDs, which are p-doped and collect electrons, the CID collects holes. Underneath the epitaxy is the substrate, which consists of highly p-doped silicon and serves as both mechanical support and common conductor for the device.

The most important features of the MOS structure are its ability to store charge indefinitely and to transfer that charge to adjacent MOS structures. When the gate is biased negative relative to the epitaxial layer, an electric field is created that extends from the gate into the epitaxy. If an electron–hole pair is created in this field, the hole will be attracted to the gate but will be unable to cross the oxide barrier. The electron will be repelled from the gate and conducted away by the substrate. This process allows holes to collect at the epitaxy/oxide interface.

2 CHARACTERIZATION

Four pre-amp-per-row CID devices were examined to assess their system gain, read noise, dark current, photometric response, quantum efficiency, and fixed pattern noise. The CID 17PPRA and the CID 38SG are commercially available devices for which prior characterizations have not been published. Although Pilon[10] examined the CID 17PPRA for use in atomic emission spectroscopy, this device has been re-examined in more detail here. The CID 38Q-A and CID 38Q-B are experimental devices that are not commercially available.

The CID 17PPRA is a 256 × 388-pixel device. Each pixel is 28 × 24 microns. The CID 38SG is a 512 × 512 device having 28 × 28 -micron pixels. Both devices are constructed using 5Ω/cm silicon. The devices are available with several epitaxial layer thicknesses: 15, 16, 20, and 28 μm.[11] The devices having 28-μm epitaxy are expected to have enhanced red sensitivity; however, no data supporting this expectation has been published. Based on the data in Hanley et al.,[12] increasing epitaxy thickness without increasing the resistivity of the silicon will probably not greatly affect the performance of the device.

The CID 38Q-A and the CID 38Q-B are based on the design of the CID 38SG, except that the Q devices are made with an all-oxide fabrication process. The structure of the devices is proprietary at present but is similar in architecture to the CID 38SG. Multiple motivations for the use of an all-oxide process exist. All-oxide amplifiers are thought to have lower noise than those made with the nitride process. In addition, the all-oxide process requires fewer process steps and for this reason is less costly.

2.1 System Gain

A CID is read using an analog-to-digital converter. The conversion results in a measurement in an arbitrary scale. This scale, graduated in arbitrary digital units (ADUs), must be calibrated further in order to relate the value in ADU to the number of photogenerated hole pairs collected in a pixel of the device. The system gain of a device relates the number of collected electrons[*] to the measured output in ADUs. Hence, measurement of the quantum efficiency of a CTD requires knowledge of the system gain.

Two methods for the measurement of system gain have been described in the literature: mean variance[13] and X-ray photon transfer.[14] The latter method relies on exposure of the device to a low flux of X-rays, usually produced by radioactive decay. The energy of the X-ray is converted into a number of electrons, which are then measured. This method requires good charge collection efficiency and low noise.

This method is difficult to implement in the present generation of CIDs, as these devices have poor collection of X-ray-generated charge. For this reason, the mean variance method must be used. This method relies on the fact that the variation of photons distributed spatially or temporally follows a Poisson distribution. For visible photons, only one charge carrier is generated for each photon. A CTD can then be exposed to an even optical field. The Poisson, or shot, noise (σ) of the carriers collected in the pixels is dependent on the number of carriers q as predicted by Equation 1. The development of mean variance presented here in Equations 1–5 follows Sims,[15] Billhorn,[16] and Sweedler et al.[6]

$$\sigma^2 = q \tag{1}$$

The measured signal, S, is related to the number of carriers collected by the system gain, G, of the device and readout electronics.

$$S = G^2 q \tag{2}$$

The measured variance, Δ^2, of the values of S is also related to the number of charge carriers.

$$\Delta^2 = G^2 q \tag{3}$$

By substituting Equation 3 into Equation 2 the standard mean variance expression is reached.

$$\Delta^2 = GS \tag{4}$$

The gain parameter G reflects the gain in all stages of the measurement process. It usually has components from both the on-chip amplifier and read out electronics and is not a fundamental property of the device.

Equation 4 implies that a plot of the set of signals measured at varying illumination levels versus the measured variance of those signals should yield a straight line whose slope is G. In real systems, there are sources of variance other than Poisson noise from incident photons. This variance from other processes is usually treated as an aggregate and is called read noise, σ_r. In scientific grade CTD systems, this noise source is independent of signal strength. The combined variance from photon shot noise and read noise is given by Equation 5.

---

[*] CIDs actually collect holes. In the characterization literature describing CIDs, collected charge is usually referred to as electrons to be consistent with the CCD literature.

$$\Delta^2 = GS + \sigma_r^2 G^2 \tag{5}$$

This predicts the relationship between the magnitude of a signal and its variance and gives a way to estimate both the system gain and the read noise of a CTD system.

The read noise, σ_r, like the system gain, reflects the overall noise characteristics of the CTD system, including the device, readout electronics, and digitization noise. As measured, it does not necessarily reflect the noise from the device alone and may vary if different controller electronics are used. However, in a well-designed system, the read noise will reflect the noise of the chip; this may be achieved by using sufficient on-chip gain such that noise added by the readout electronics is negligible.

Practical application of the mean variance Equation 5 requires careful isolation of variance from read noise and shot noise from potential confounding factors such as fixed pattern noise and row crosstalk. The present generation of CIDs has a large fixed pattern noise (see Section 2.5). This is not noise in the strictest sense. Fixed pattern noise in CIDs manifests as an offset that varies from pixel to pixel. The measured value from a pixel, M_i, is a sum of the pixel offset, P_i, and the signal due to the presence of charge carriers, S_i.

$$M_i = P_i + S_i \tag{6}$$

When performing the mean variance experiment based on the spatial distribution of photons, the measured signals from a large number of pixels are averaged. In a CID, large pattern noise from the device results in the mean variance plot being compromised by the pattern noise. The reason for this result is that mean signal and variance reflects the fixed pattern noise rather than the collected photons, as shown in Equations 7 and 8.

$$\overline{M} = \frac{\sum_n (P_i + S_i)}{n} \tag{7}$$

$$\Delta^2 = \frac{\sum_n \left((P_i + S_i) - \overline{M}\right)^2}{n - 1} \tag{8}$$

In Equation 8, the measured variance is compromised by the values of P_i. Sims[15] described an exposure sequence that removes fixed pattern noise from a mean variance experiment by computing the variance on the difference image between a pair of matched exposures. The mean of the difference image is zero and the offsets, P_i, cancel. Because variance adds when taking a difference, the resulting variance, Δ^2_m, is twice that of Equation 5.

$$\Delta^2_m = 2\Delta^2 \tag{9}$$

A further problem with the implementation of a good mean variance experiment is assuring that the paired exposures have similar illumination levels. In all data reported here, exposures were checked to ensure that the illuminations matched. A variety of diagnostics were also checked to ensure that the distributions were nearly normal by computing the skew and curtosis of the values in the difference image. A significantly skewed distribution or one with inappropriate curtosis indicates a measurement problem, and the process was repeated. These diagnostics were developed first for use with the CID38Q-A, which had readout problems resulting in randomly appearing bad rows in images. In other devices, when a strip five pixels wide is removed from the edges of the subarrays used for measurement, the distributions were well behaved and good results were readily obtained.

A final problem in implementing the mean-variance method was encountered because of the non-linear response curve of all the CIDs examined when evenly illuminated over large areas of the device. The mean-variance method assumes a linear response to collected charge. Adaptation of the method to a non-linear system was attempted but was found to be intractable. Instead, illumination was restricted to low levels where the response closely approximates a straight line.

2.1.1 CID 17PPRA Device System Gain. A typical mean variance plot for a CID 17PPRA appears in Figure 1. Prior to this work, there were no reports of the variation expected to be present in a set of CID devices or in replicate measurements made on a single device.

A set of devices was studied using a standard protocol. The system gain was measured at two pixel frequencies, 16.1 and 9.43 kHz. In all cases, the exposures were limited such that the maximum exposure level in the mean variance experiment was less than 100,000 carriers. This level corresponds to the lower 10% of the CID's well capacity. All mean variance experiments contained 50 exposure levels. These data are tabulated in Table 1. These data allow an evaluation of the reproducibility of the mean-variance measurements within a single device and between different devices. All the data shown in Table 1 were taken using a single board set.

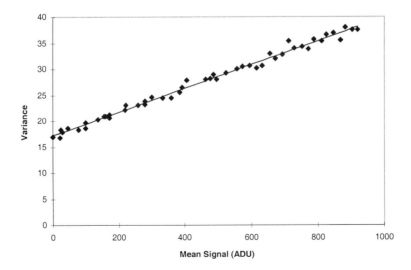

Figure 1 *Mean variance plot for CID 17PPRA device number 16. The system gain obtained from this plot is 43.8 carriers/ADU. Data taken using a CID gain parameter of 130. This corresponds to a 16.1 kHz pixel frequency.*

Table 1 *System Gain Data for 4 CID 17PPRA Devices*

Measurements at CID Gain Parameter 130 (16.1 kHz)

Device					mean ± 1 sd
#15	46.69	45.75	47.71	46.23	46.60±0.84
#16	43.84	45.58	45.59	46.40	45.35±1.08
#17	45.58	45.30	46.01	45.73	45.66±0.30
#19	45.78	44.02	45.25	45.84	45.2±0.8

Measurements at CID Gain Parameter 240 (9.43 kHz)

Device					mean ± 1 sd
#15	27.61	27.18	27.71	29.23	27.93±0.90
#16	26.50	24.81	25.48	26.15	25.73±0.75
#17	25.63	25.87	24.72	25.77	25.50±0.53
#19	24.76	26.08	26.72	24.65	25.55±1.00

The data indicate that the exposure sequence used gives reasonable reproducibility. Only a few devices need to be examined, and the system gain shows little variation within a single family of devices. The data taken at 9.43 kHz has a one standard deviation error of 4.9% and the corresponding error at 16.1 kHz is 2%. These tolerances

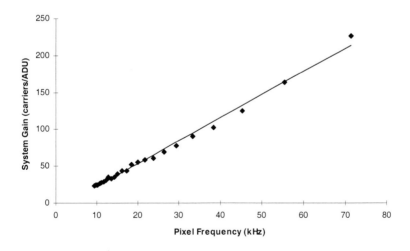

Figure 2 *Plot of the relationship between pixel frequency and measured system gain for a CID 17PPRA device. The line corresponds to the best-fit straight line to the data. In an ideal system, the data should match a straight line. At low frequency, the correspondence with the ideal is moderately good. At higher frequency, some curvature is apparent.*

are important, as the measurement of the read noise and quantum efficiency relies on the system gain. Therefore, both these quantities should be assumed to have at least 2–5% error. The CID gain parameter is an arbitrary index that controls the length of integration on a correlated double sample. As the time of integration increases, the system gain in carriers/ADU decreases. It is desirable to optimize the CID camera system by minimizing the read noise. To evaluate this, knowledge of the system gain as a function of CID gain parameter (pixel frequency) must be known. A series of system gain measurements, taken at intervals of 10 gain parameter units, were made in a number of devices. A representative plot of such a set of measurements is shown is Figure 2.

In the CID 17PPRA, this type of plot was reasonably reproducible between devices, and good measurements were typically possible from 9–80 kHz. Some instability was observed at CID gain parameters less than 20 (55 kHz and above).

2.1.2 CID 38Q-A Device System Gain. Two CID 38Q devices were studied. The CID 38Q-A was a first-generation all-oxide prototype device that was initially expected to have lower read noise, higher radiation tolerance, and lower production cost. Only two devices were available for characterization. System gain plots for one of these devices appear in Figure 3. The CID 38Q-A device performed poorly, and was generally unsuitable for scientific purposes.

2.1.3 CID 38Q-B Device System Gain. The CID38Q-B is a second-generation all-oxide process CID. A variety of modifications were made to the original device to correct some of the problems of the CID38Q-A. A system gain plot of this device appears as Figure 4.

2.1.4 CID 38SG Device System Gain. The CID 38SG is a nitride process device. This device has been commercially available for several years, and its design forms much of the basis for the Q device. A system plot for this device appears in Figure 5.

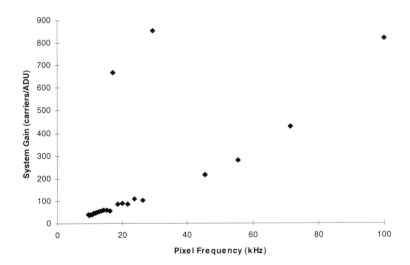

Figure 3 *Plot of the relationship between pixel frequency and measured system gain for a CID 38Q-A device. The performance of the device was generally poor. The worst two outliers were removed. The remaining "poor" data points have been included and are an indication of the performance of this device.*

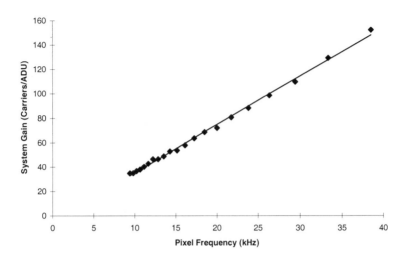

Figure 4 *Plot of the relationship between pixel frequency and measured system gain for a CID 38Q-B device. The performance of the device was adequate.*

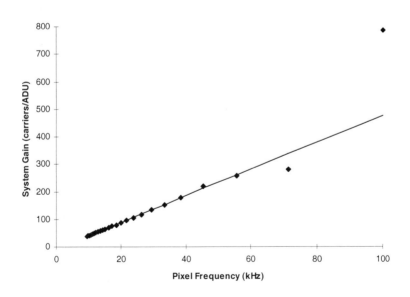

Figure 5 *Plot of the relationship between pixel frequency and measured system gain for a CID 38SG device. The line corresponds to the best-fit straight line to the data. The two highest pixel frequencies have been eliminated from the best-fit line.*

2.2 Read Noise

The read noise of a device is the sum of all signal independent noise except pattern noise and dark current noise. It is commonly used as a measure of the ability of a device to detect low-level signals. In the CID, read noise is generally higher than it is in CCDs, because the magnitude of the output signal from a CID is limited by the total capacitance of the sense circuitry. This limitation results in output signals of 30–50 nV/carrier before amplification, about two orders of magnitude lower than CCD output[17], and as such is more susceptible to degradation. The sensitivity decreases with device capacitance (size). Models of pre-amp per row CID performance consider row pre-amplifier noise as the greatest source of noise in the device.[18] In practice, other sources of noise are of concern. The ability to have output correspond to the signal from the device requires careful design of the readout electronics so that subsequent processing does not degrade the output signal. Experience has shown that interference from nearby computer monitors and electric fields seriously degrades the signal. To achieve good measurements from the CID camera systems, it is imperative to shield the readout electronics relative to other laboratory equipment.

The flexibility of the CID makes it possible to assess the read noise in several different ways. Further, the different readout sequences cannot be assumed to give the same level of measured noise. In particular, it was unclear whether readout sequences incorporating injection (charge clearing) would result in measured read noise consistent with sequences without injection.

In the CID literature, noise is an estimate of the error in a measured signal. Two statistically distinct quantities are used interchangeably and reported as read noise. Reported read noise in CIDs may be either the standard deviation of series of measurements or the estimate of the error in the mean value of a series of measurements. Since CIDs can be read repeatedly without destroying the accumulated charge, the expected error associated with the mean signal decreases. However, it should be noted that the standard deviation of a series of measurements does not decrease with the number of measurements made. A better estimate of the standard deviation is obtained but the magnitude does not decrease. The estimate of the mean value of the distribution also improves. The error in the mean value is sometimes referred to as the error of the mean or the standard error of the mean. In this chapter, to avoid confusion, read noise will be used only to describe the standard deviation of a set of measurements made with a device, and, when multiple measurements of a single packet of accumulated charge are made, the subsequent error will be referred to as read error. The CID literature does not normally make this distinction and it is sometimes difficult to ascertain the conditions of measurement responsible for reported values of read noise.

2.2.1 Read Noise Method A: Intercept. Equation 2.5 may be used to estimate the read noise of a device. In this method, the best-fit line to a mean-variance plot is computed. The slope gives the system gain, G, and the intercept is $G^2\sigma_r^2$. Solving for σ_r results in a value for read noise. This method is problematic, as it is influenced by the photometric response of the device. Depending on the readout sequence and the quality of the device, a "foot" is often observed in CID mean variance plots. The foot is a region of low photometric response in which charge carriers accumulate in trap sites and other defects rather than in the potential well of the pixel. These trapped carriers are not measured, which results in a decrease in the measured signal. Further, the photometric response of the CIDs examined here shows curvature at high values of mean and variance, which can result in over-estimation of the device read noise. In the case of a "foot" in the plot, the read noise of the device is underestimated. In the case of curvature at the high end of the plot, the read noise is over-estimated.

2.2.2 Read Noise Method B: Paired Subarray Reads. A set of paired reads may be used to measure the read noise of a device. The device is cleared of all charge and a set of paired reads of a subarray on the device is measured. The difference between the two measurements is computed pixel by pixel to remove fixed pattern noise. The standard deviation of the difference is computed and divided by two (see Equation 2.9). The standard deviation in ADU is multiplied by the system gain to obtain the read noise in carriers.

$$\sigma_r = \frac{G}{2} \sqrt{\frac{\sum_n \left(S_i - \bar{S} \right)}{n-1}} \tag{10}$$

In (10), S_i is the difference signal from the ith pixel in the subarray and $\bar{S}$ is the mean of the difference subarray.

Two variations of this measurement were carried out. In the first variation, the device was cleared and the same subarray was read twice to obtain the data. This type of measurement was termed read noise. In the second variation, the device was cleared of all charge, the selected subarray read, the charge was cleared from the device using the CID's inject feature, and the subarray was then read a second time. The result of this type of measurement was termed inject noise. The purpose of the latter measurement was to determine whether the process of injection introduced an additional source of noise to the measurement process.

2.2.3 Read Noise Method C: Standard Deviation of Rereads. In an approach unique to the CID, a set of rereads from a single pixel is used to estimate the read noise of a device. A set of reads from a pixel is made. The standard deviation is measured and then converted to charge carriers.

$$\sigma_r = G \sqrt{\frac{\sum_n \left(S_i - \bar{S} \right)}{n-1}} \tag{11}$$

This expression is very similar to (10) except that the factor of two has been eliminated, and S_i represents the ith measurement of the signal in a single pixel. This expression can be used to establish whether the noise is uniform over the surface of a CID.

2.2.4 Read Noise Method D: Moving Average in a Series of Rereads. In most noise measurements, it is assumed that noise is "white", or that the time scale of a series of measurements is such that unwanted signals appear to be "white." In practice, a system might not be well enough behaved to allow this assumption to be made. An example of a confounding noise source might be the case of a 60 Hz signal contaminating a measurement. Such noise sources are common in laboratories using 60 Hz power sources. If a series of 20 measurements are made of such a system, in a burst taken at 20 kHz, a non-representative sample of the 60 Hz noise is obtained. The subsequent calculation of the read noise will give a spuriously low value. If the samples are made at randomly spaced time intervals over several minutes, a more representative measurement will be obtained and the read noise will be accurate.

To assess the time-varying properties of the read noise, a number of experimental approaches were used. First a 20-point moving average was computed over a series of 1000 measurements. The standard deviation of the moving average was computed and converted to charge carriers and the results plotted as a function of position in the array. Plots of this type showed periodic trends and non-random behavior.

Fourier power spectra were also analyzed for this type of data. Such spectra showed expected 1/f (flicker noise) and constant components (white noise). They also contained large components not attributable to either of these noise sources. The frequency of these noise sources varied depending on the pixel frequency used. It is not clear what the source of these signals was, or if they were perhaps being aliased in from higher frequency signals. It is clear that the readout electronics were a major source of read noise, and that it is doubtful the present generation of readout electronics allows the inherent read noise of a CID device to be assessed. The values reported here should be assumed to be high.

2.2.5 Read Noise of CID 17PPRA. The values for read noise of a series of CID 17PPRA devices appear in Table 2. As can be seen, measured read noise depends on the readout electronics used for measurement.

Table 2 *Value for Read Noise Taken at 9.4 kHz*

CID 17PPRA	M-V[*] Read Noise	Read Noise	Inject Noise	M-A Noise[†]
# 15	251	248	230	NA
(B3FPA1)[‡]	193	181	183	NA
#16	198	NA	NA	NA
(B3FPA1)				
#17	206	204	195	NA
(B3FPA1)				
#19	198	201	187	NA
(B3FPA1)				
#10	158	148	143	NA
(B1FPA2)				
#10	105[§]	141	141	140
(B1FPA2)				
#11	115[§]	139	140	139
(B3FPA2)				

All values in the table are in charge carriers. The characters in parentheses refer to the controller and FPA board used. A total of three controller boards and three FPA boards were examined. This table summarizes the most comparable experiments.

The data in Table 2 indicate that the intercept method is indeed inaccurate for cases where a foot is present in the photometric response curve. Where overlapping data for the other three methods are available, agreement is quite good. As seen by the first entry in the table, the position of the camera relative to computer monitors and other laboratory

---

[*] M-V Read Noise is a value of read noise obtained from the intercept of the mean variance plot.

[†] M-A noise is moving average noise.

[‡] The first entry in the table was taken with a computer monitor close to the camera head. The second value is the same device in the same board set with the monitor at a distance.

[§] This value is due to a foot in the mean variance curve. See text for discussion of the effect of a foot in the mean variance plot.

equipment can cause the read noise to increase. Based on these data the read noise of the CID 17PPRA device can be considered to be less than 140 e-. This level of performance, however, may not be obtainable in day-to-day measurements, as it depends on the particular conditions and read out electronics used.

The read noise was measured using the intercept method over the range of pixel frequencies available to the camera system in order to determine the optimal frequency for measurement. Although many devices were examined, Figure 6 is representative of the trends observed. Best measurements are made between 12–20 kHz.

The read noise of a CID 17PPRA was measured using the moving average method. The noise measured as a function of the position in the array of reads is plotted in Figure 7. The average of a set of measurements, such as those plotted in Figure 7, agreed with those taken by other measurements. The range was over 150 carriers, indicating that values of read noise should be treated with some caution and that a large population is necessary to obtain an accurate measurement of the output of a device.

2.2.6 Read Noise of CID 38Q-A and 38Q-B. The read noise of the two all-oxide process devices was measured using the intercept method. These data are plotted in Figure 8. The mean variance plots of the latter device showed a wide foot.

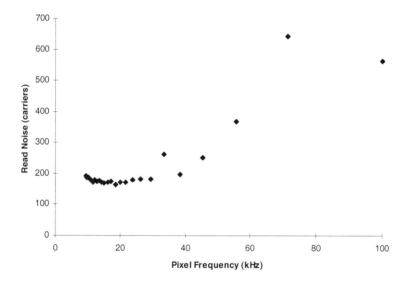

Figure 6 *All the data in the figure were taken using the same device, using the intercept of a mean variance plot. The increased scatter at high frequency was common to all the devices observed. The read noise could not be reliably measured above approximately 30 kHz. Optimal measurements could be made between 12–20 kHz. It should be noted that there is no reason to use the device at pixel frequencies below 12 kHz. The full dynamic range is unavailable because of saturation of the output electronics, and the noise is proportionately higher.*

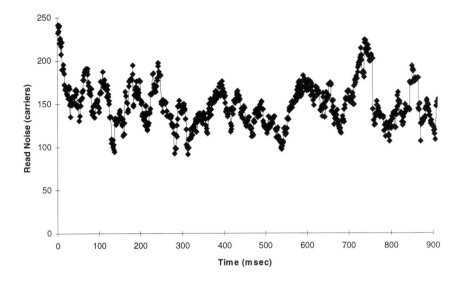

Figure 7 *The noise computed from a 20-pt moving average using a CID 17PPRA device.*

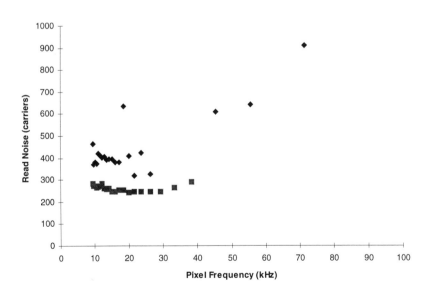

Figure 8 *Read noise of the CID 38Q-A and CID 38Q-B measured at varying pixel frequency. The filled diamonds correspond to the first-generation CID 38Q-A device. The filled squares are for the CID 38Q-B.*

Table 3 *Controller Board Comparison*

Board Set	Read noise (carriers)
#1	171-194
#2	192-220
#3	226-240

2.2.7 Read Noise of CID 38SG. Three controller board sets were tested using the same FPA board and a CID 38SG device. These data are presented in Table 3. One of the sets of controller boards worked noticeably better than the others and was used for subsequent measurements of read noise.

A plot of the read noise of a CID 38SG obtained using the intercept method is shown in Figure 9. The device gives a read noise of about 180 carriers over a wide range of frequencies.

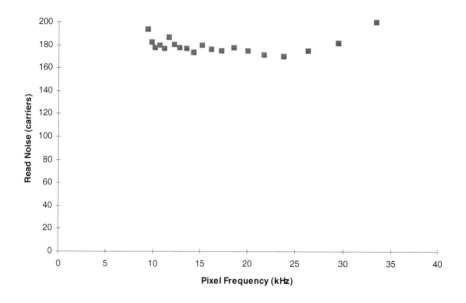

Figure 9 *Read noise plot for the CID 38SG as a function of pixel frequency.*

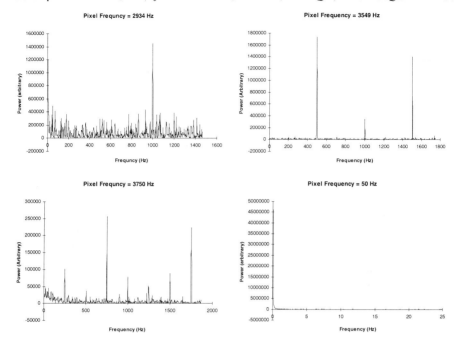

Figure 10 *Power spectra of a series of rereads from single CID 38SG pixels taken at varying pixel frequencies. The data taken at 50 Hz shows a dominant 1/f component. The data taken at higher frequencies showed strong non-random components that do not correspond to either 1/f or "white" noise. The change in position of the peaks in the power spectra with pixel frequency suggests an aliased high-frequency signal or time-dependent error in the timing circuitry.*

In the study of the CID 17PPRA, it was noticed that there was some non-random behavior in a series of non-destructive reads. To further assess this behavior, sets of measurements were made at a variety of different pixel frequencies. The power spectra of the noise in the series of reads were measured and the data plotted as a function of frequency. These data are shown in Figure 10. As can be seen, depending on the frequency of acquisition, strong, non-random components to the power spectra are observed. The source of this non-random noise was never found but is thought to be an artifact from the readout electronics.

On the advice of the manufacturer, the resistance in a late-stage amplifier was decreased to see whether a reduction in the read noise could be observed. The read out electronics are illustrated in Figure 11. The circuit modification consisted of adding a 1.5 kΩ resistor in parallel with the 3.01 kΩ resistor labeled R205. This modification was predicted to increase system gain by a factor of three. The manufacturer was interested in the effect this modification would have on the read noise of the camera system. Table 4 shows the results of this modification.

Table 4 *Effect of Gain Boost on System Gain and Read Noise*

Condition (20 kHz pixel frequency)	System Gain (Carriers/ADU)	Read Noise (Carriers)
No Gain Increase	81.1	180
3X Gain Boost	26.9	200

A slight increase in readout noise was observed following the modification of the readout electronics. In other words, the system performed slightly worse than if the digital values reported by the readout electronics were multiplied by an arbitrary factor of three. It is of note that the read noise is roughly comparable to those made at 10 kHz, except that these measurements are made twice as fast. In comparison to those measurements, the overall system performance has improved by a factor of 1.41.

2.3 Read Error

As discussed in Section 2.3, the ability to repeatedly read a single CID pixel provides a reduction in the error associated with a measured signal. Others have described this capability.[10,15,16] In the prior literature, this process has been referred to as a reduction of read noise. Using the nomenclature of Section 2.2, this is a reduction in the error of a mean. It should also be noted that the ability to reduce the error in a measurement assumes that individual reads of a given pixel are dominated by white rather than flicker noise and that the reads are uncorrelated.[15] As a result of these restrictions, and the behavior demonstrated in Figure 10, reduction of error using this technique should be treated with caution. The noise is not white when measured using the current generation of readout electronics. It should also be noted that non-white signals could be reduced, provided that the time scale and sampling scheme are such that the signals appear to be white. If these conditions are not met, then the expected error reduction will not be observed.

The standard deviation for a series of measurements is given by Equation (12).

$$\sigma = \sqrt{\frac{\sum_n \left(S_i - \bar{S}\right)}{n-1}} \tag{12}$$

where $\bar{S}$ is the mean signal of the n measurements and S_i is the *i*th measurement. As more measurements are made, better estimates of the mean and standard deviation are obtained. The error in the mean value is related to the standard deviation by Equation (13).

$$s = \frac{\sigma}{\sqrt{n}} \tag{13}$$

This equation predicts that error in a series of measurements decreases as the number of observations increases.

In previous work,[15,10] the observed error typically exceeded values predicted by (13). Sims et al. attributed this excess to digitization noise and to the bandwidth of the readout electronics.[19] The behavior of the new generation of devices was of interest, both to see whether the theoretical noise reduction could be achieved and to optimize the readout.

Plots of the reduction in read error as a function of the number of non-destructive reads are presented in Figure 12. The data were taken at approximately 200 Hz. Near-theoretical-error reduction was observed.

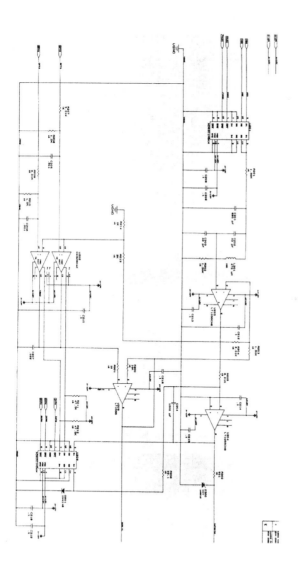

Figure 11 *Output amplifier circuitry*

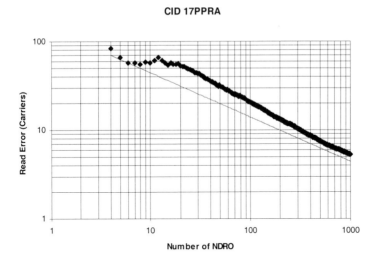

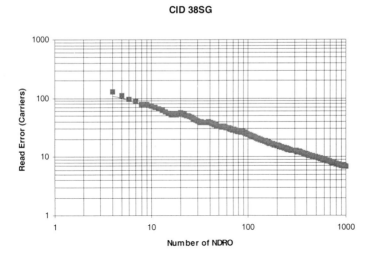

Figure 12 *Decrease in measured signal error as a function of the number of non-destructive readouts. The connected line corresponds to the theoretically predicted reduction in error. Correspondence is particularly good for the CID 38SG device. In the other plot, a few reads of high variance at the beginning of the measurements result in an offset from the theoretical curve (solid line).*

2.4 Pattern Noise

In Section 2.2, pattern noise is briefly discussed. All charge-transfer devices suffer from some level of pattern noise from pixel-to-pixel variations in sensitivity, various types of defects, processing limitations, and other sources of random variation in pixel output. Pre-amp-per-row CIDs are particularly prone to this type of noise.

The pattern noise in CIDs comes from a variety of sources: defects introduced during manufacturing, small variations in row amplifier offsets, and any other process that causes

variation in row and column capacitance. Pattern noise is not "noise" in the truest sense of the word. It is fixed variation and can be removed completely by subtracting off a bias image. The total amount of variance in a series of images taken at zero exposure can be treated as a sum of individual variances. Using this approach, a model can be constructed based on variance from rows, columns, pixels and reads.

$$\sigma^2_{total} = \sigma^2_{col} + \sigma^2_{row} + \sigma^2_{pixel} + \sigma^2_{read} \tag{14}$$

The "fixed pattern" noise is due to the components from rows, columns, and pixels. To separate the components, the read noise is first measured using the paired subarray technique. The total noise is measured. The column means are subtracted from each value in the corresponding column of the subarray, and the remaining variance is computed.

$$\sigma^2_{total} - \sigma^2_{col} = \sigma^2_{row} + \sigma^2_{pixel} + \sigma^2_{read} \tag{15}$$

The row means are then subtracted from each value in the corresponding row of the subarray. This leaves pixel and read noise. The read noise is known in advance, which allows all individual variances to be solved. Table 5 gives the results of these measurements.

The major source of pattern noise in each of these devices is different. In the CID 17PPRA, a first generation pre-amp-per-row device, far greater overall pattern noise is seen, and the greatest source is column-to-column variation. In the second generation CID 38SG, a great improvement in both row-to-row and column-to-column variation was realized. The pixel-to-pixel variations are somewhat larger in the CID 38SG. We believe this modest increase is due to the larger pixels in the CID 38SG. The ratio of pixel noise for the two devices (0.83) closely matches the ratio of pixel areas (0.85).

Table 5 *Pattern Noise*

Noise Source	CID38SG[*]	CID 17PPRA[†]
Total Noise (carriers)	4827	7326
Row Noise (carriers)	988	3714
Column Noise (carriers)	1241	5055
Pixel Noise (carriers)	4554	3782
Read Noise (carriers)[‡]	220	133

[*] CID 38SG device measurements were made at 20 kHz.

[†] CID 17PPRA measurements were made at 13.5 kHz.

[‡] As noted elsewhere in the text, the measured values of read noise depend on the particular readout electronics used. The read noise reported in this table reflects the camera control unit used to make the measurements.

2.5 Conclusions for System Gain and Noise

There has been considerable speculation about the limiting source of noise in the CID. Most reports in the literature refer to the high capacitance of the sensing circuitry as the limitation in making low noise measurements with a CID. Recently, amplifier noise on the device has been used to model CID read noise.[18] Our results indicate that, in practice, amplifier noise is not dominant. Rather, the read noise of the systems tested here seems to be limited by the readout electronics and pre-amplifier gain. The CID 17PPRA showed lower noise than did the CID 38SG. This can be explained in the following way: the ratio of read noise (1.28) is close to the ratio of row length (1.32) and to the ratio of system gain. The row length is a rough indicator of the sense capacitance. As the capacitance increases, the output signal decreases, which makes the system gain of the larger device (CID 38SG) smaller. The smaller output signal is then degraded to a greater extent by subsequent readout electronics. For read noise to be equivalent to that of the CID 17PPRA, a higher gain preamplifier must be incorporated into the chip design. Future devices should strive to increase the level of gain on the chip. This will result in larger signals leaving the device and will reduce the effect of post-amplification contaminating signals. It is also apparent that the current camera controller electronics are not well optimized for the pre-amp-per-row architecture.

2.6 Photometric Response Function

The photometric response function of a device defines the signal level in terms of the incident radiation on the device. CIDs have been shown to exhibit non-linear behavior,[10,19,16] i.e. that CID response is not a simple linear function of the incident light intensity. Early CIDs showed an initial "foot" or "fat zero" in the response curve.[19] Outside of the foot region, Sims observed nearly linear behavior in the CID11B. Bilhorn,[16] working with the CID17, noted that the device gave a non-linear response after the foot region was saturated. The effect could be corrected using a 2nd degree polynomial. Bilhorn attributed this behavior to *diminished* capacitance due to the large amount of charge transferred to the row electrode. Pilon[10] noted in his initial work on the CID17PPRA device that the region following the "fat zero" also was non-linear. True[20] also observed this effect while working with the CID 38SG. The effect is strongest when the device is evenly illuminated. It has been speculated[16,10] that the non-linearity was due to decreasing row and column capacitance as the device approached saturation. The change in voltage measured by the amplifier depends on the capacitance, as shown in Equation 16.

$$dV = \frac{dQ}{C} \tag{16}$$

Equation 2.16 suggests that as C decreases, dV will increase. This predicts a progressively increasing signal as more photogenerated charge is added to the potential well. A derivative plot should show an upward slope.

Figure 13 shows a photometric response plot for a CID 17PPRA and its derivative. The slope of the derivative plot is the opposite of what would be expected if the capacitance of the device is decreasing.

Previous work by Sims[19] showed a nearly linear response curve over much of the photometric response curve for the CID 11. Pilon[10] found that all CID 17 generation devices showed non-linearity, both with and without row-pre-amplifiers. True[20] has shown that similar non-linear behavior is seen in the CID 38SG and that output non-linearity is related to another phenomenon called *row crosstalk*, a relationship also suggested in Bilhorn.[16] If a CID 38SG is evenly illuminated, the extent of non-linearity increases. If only a small spot, a few pixels in size, is imaged, the response more closely approximates a straight line.

The existence of CIDs of a generation preceding the current pre-amp-per-row devices, one with good linearity, another with a non-linear response, suggests that the problem is in the readout electronics. Non-linearity in the pre-amp-per-row devices is more problematic. First, the row pre-amps each consist of a single FET. FETs are not inherently linear components, and, unless operated correctly, produce a non-linear response. As a result, the observed non-linearity in the CID 38SG and CID 17PPRA may be attributable to a combination of device architecture and readout electronics. Carbone[21] has indicated that the problem is opposite to that in prior speculation by Pilon and Bilhorn. Carbone suggests that the problem is related to row collection of charge in the present electronics. Pilon[10] originally adopted row collection when he noticed that its use improved device performance. Based on this work, row collection was adopted by CIDTEC and incorporated into the SCICAM hardware and software used in these studies. Rather than decreasing the row capacitance as the device well capacity is being reached, the capacitance increases, resulting in the behavior observed.

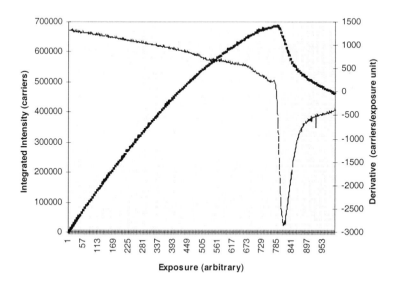

Figure 13 *Photometric response curve for the CID 17PPRA. Heavy line is integrated intensity. Light line is a first derivative plot for the integrated intensity data. As can be seen, the device is not linear over any portion of its range.*

Sims et al.[22] described a method for cross-talk correction to enhance photometric response. True[20] has shown that injection of areas outside the region of interest improves linearity. Correcting output nonlinearity is possible provided that the exposure conditions of the experiment and the calibration procedure are similar. Such a procedure was adopted by Bilhorn[16] and was also suggested by Pilon[10] using a second order polynomial. The data in Figure 13 may be fit to a second order polynomial up to about 80% of the maximum signal. To fit the range up to 90% of the maximum signal required a 4th order polynomial. None of these procedures are entirely satisfactory. In practice, such linearization procedures are of limited utility, because the output depends upon whether the entire device is illuminated or only a small portion thereof. For practical applications, it is imperative to verify that, under the conditions of the experiment, suitable linearity is observed. Additional discussion of CID linearity may be found in True[20] and Hanley.[23]

The non-linear output of the device makes calculation of the full well capacity problematic. Pilon[10] measured the well capacity of the CID 17PPRA to be 9.6×10^5 carriers. No correction for linearity was performed. The data in Figure 13 indicate a well capacity of 6.9×10^5 carriers if no correction is applied. If a correction is applied, the well capacity is 1.0×10^6 carriers.

2.7 Quantum Efficiency

Quantum efficiency is a measure of a device's ability to detect photons. This term is defined variously in the literature, but is usually described as the ratio of incident photons to detected photons expressed as a percent. However, some workers (cf.: Barnard[24]) have broadened the definition to include the ratio of incident photons to detected charge carriers. This quantity is more appropriately called quantum yield. Under the usual definition, the maximum measured quantum efficiency should never exceed 100%. This definition is adopted here, because the broader definition is absurd in the X-ray region, where the number of photoelectrons generated from a single photon can exceed several thousand.

In the optical range, quantum efficiency can be measured through the use of a calibrated light source. Here, a NIST traceable photodiode (EG&G UV 444BQ) was used to calibrate the output from a monochromator. The number of photons passing through the exit aperture was measured using the photodiode. The number measured using the photodiode was compared to the number of carriers generated in the CID. Quantum efficiency curves for the CID 17PPRA, the CID 38SG, and the CID 38Q-A and CID 38Q-B are shown in Figure 14. For these measurements, some devices were coated with Metachrome II to enhance UV performance. This fluorescent substance absorbs UV light and re-emits it at longer wavelengths.

The CID 17PPRA was previously examined and reported to have a QE maximum of 82% at 550 nm, almost double the value of 45% reported here. The value of 82% at 550 nm is also inconsistent with other measurements made on the CID17. Given the known reflection losses off the SiO_2 surface layer and Al straps, the previous results are probably high.

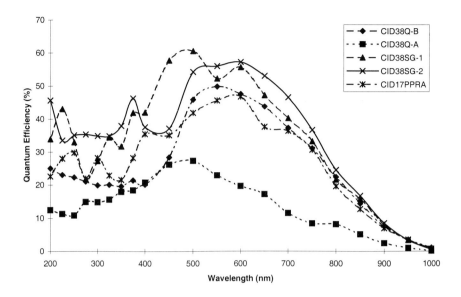

Figure 14 *Quantum efficiency curves for the CID 38-A, CID 38-B, CID 38SG, and CID 17PPRA. The CID 38Q-B and the device labeled CID 38SG-2 have Metachrome II coatings to enhance UV response.*

2.8 Dark Current

Dark current can be a significant source of temporal noise in measurements made with a charge transfer device. In previous measurement with CIDs, liquid nitrogen cooling was used to reduce the magnitude of dark current. When cooled to liquid nitrogen temperatures, CIDs show negligible dark current, and could not be measured with any accuracy.[23]

When cooled to liquid nitrogen temperature, dark charge generation is exceedingly slow. Even if 1000 re-reads are used, 10–20 carriers/per pixel must be generated before a measurement of reasonable accuracy can be obtained. This can take days or weeks.

There are, however, a number of disadvantages to liquid nitrogen cooling. Liquid nitrogen is expensive, requires high vacuum, and requires frequent filling if measurements are to be made over several days. Furthermore, the rate of charge recombination is slow enough that it can be difficult to clear accumulated charge from devices cooled to liquid nitrogen temperatures. Great care must be exercised when using CIDs at liquid nitrogen temperatures or image ghosts may be observed in acquired images. It may be preferable to operate the device at a higher temperature. For this reason, it is desirable to evaluate the amount of dark current generated as a function of the temperature of the device. With such data, the most convenient cooling system for a particular experiment can be selected.

A CID 38SG and a CID 38Q-B were supplied by Thermo Jarrell Ash Corporation and installed in a refrigerated housing with a variable temperature refrigeration unit. Using this system the dark current was measured over the range from -100 to 0 °C.

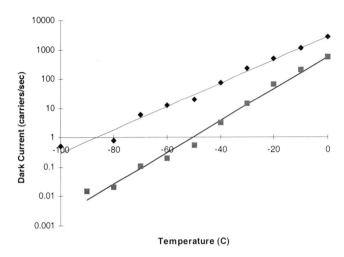

Figure 15 *Dark current for the CID 38Q-B and the CID 38SG. The filled squares are for the CID 38SG. The solid diamonds are the CID 38Q-B. The CID 38Q-B device had between 4 and 40 times greater dark current depending on the temperature of measurement.*

These data are plotted in Figure 15. The CID 38SG device showed dark current values similar to high quality CCD devices (cf.: Strunk et al.[25]). The CID 38Q-B had a higher dark current at all temperatures of measurement and the rate of decrease with temperature was lower than seen in the CID 38SG. No CID 17PPRA devices were available for the cryogenic assembly which is used with this refrigeration system, so comparable measurements could not be made.

2.9 Hysteresis Effects

Hysteresis effects in CIDs have been described by Sims.[19] The hysteresis reported by Sims in non-pre-amp-per-row CIDs is due to release of trapped charge from slow trap sites at the SiO_2/Si interface. A previously undescribed type of hysteresis was observed during imaging studies using the CID 38SG at liquid nitrogen temperatures. This hysteresis is observed when the device is turned on while at liquid nitrogen temperature. The first image taken with the device will appear normal. When this image is cleared a residual "ghost" image will appear. Repeated attempts to clear the charge are unsuccessful using the normal injection procedure. If the device is turned off, then turned back on the image clears completely. This type of hysteresis has been observed by others.[26,27]

This hysteresis is due to the way in which charge is collected and stored in a CID. The charge is collected in a potential well beneath a pair of crossed electrodes. During charge collection, the electrodes are biased negatively with respect to the device epitaxy. During injection, the bias voltage is set close to the epitaxial layer. During this process the potential difference between the epitaxial layer and the inject level may vary from 0.0

V to a few tenths of a volt. Values close to zero are not recommended because a larger fat zero or foot region is seen. This is due to de-population of trap sites. If the inject level is set slightly lower than the epitaxy, a small potential well still exists. At liquid nitrogen temperature, there is not enough thermal energy present to inject all of the stored charge. Some of it stays in the well giving rise to a "ghost" image. This type of image ghost is rarely seen in devices operated in refrigerated housing or at room temperature. Under these conditions, carrier generation in the dark fills the residual well and thermal energy allows the trapped charge to migrate more rapidly.

Injection voltages need to be chosen carefully and for best results a bias image should be collected. A bias image must be taken after the previous image has been injected. This allows the "ghost" image to be removed. Subtraction of a bias image also raises the effective read noise in an image. Another approach to this hysteresis is to expose the entire device to a saturating light level and then inject. This leaves the potential wells filled evenly after injection.

A second type of hysteresis was observed during imaging of Laue patterns: a ghost image that appeared even after proper collection of bias images. This ghost appeared intermittently during data collection. Although the exact cause is unknown, it is believed to be from partial warming of the camera system during long exposures. Under these conditions, the charge trap sites de-populate based on their level of prior saturation. This hysteresis is probably of the type reported by Sims.[19] An image with this type of hysteresis is shown in Figure 16.

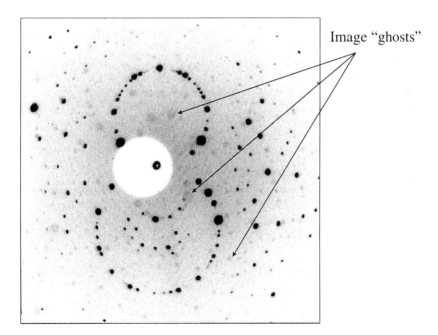

Image "ghosts"

Figure 16 *Laue image taken with a CID 38SG device using liquid nitrogen cooling. A ghost image from the previous exposure is clearly observed. Full bias subtraction was performed. The ghost image is believed to be due to release of charge from trap sites during image acquisition.*

3 SUMMARY AND CONCLUSIONS

The system gain of four pre-amp-per-row CIDs has been measured over a wide range of pixel frequencies using the method of mean variance. The exposure sequence, first proposed by Sims,[15] for removing variance due to fixed pattern noise from other sources of variance has been implemented. Further enhancements were made to render the method more robust. A series of similar devices have been examined to evaluate the reproducibility of the method of mean variance and the amount of device-to-device variation.

These studies indicate that, for CIDs, the method of mean variance is subject to approximately 2–5% error, depending on the pixel frequency used. Prior to this work, similar data on the reliability of the method with CIDs had not appeared in the literature.

The read noise of the four devices was assessed using four different methods. Of these, only one, method A (the intercept method), has been previously described. Methods similar to method B (paired subarray reads) have been described in the literature for CCDs, but have not appeared previously for CIDs. Estimates of the read noise of a device using a set of non-destructive readouts (NDROs) in either fixed standard deviation or based on the standard deviation of a moving average have not been previously reported. Using these methods, good agreement was obtained when comparable experiments were analyzed. A strong dependence on individual readout electronics was observed. Hence, the measured values are more appropriately termed "system read noise" rather than "device read noise". When optimal readout electronics were used, the measured system read noise for the CID 38SG (<180 carriers/read) and the CID 17PPRA (< 140 carriers/read) are lower than any previously measured values for CIDs. It is worthy of note that the true device read noise of the pre-amp-per-row generation of CIDs remains an open question. The values reported here should be considered upper bounds only.

Analysis of carefully timed rereads indicates the presence of non-random signals contaminating measurements made with a CID 38SG, due to non-optimal readout electronics and low inherent gain in the devices themselves. Adding gain at the last stage of amplification of a CID 38SG device resulted in theoretically predicted system gain increase. This gain boost resulted in an increase in system read noise instead of the expected decrease.

Prior reports of CID read noise in the literature have used the standard deviation and the standard error of the mean value interchangeably when reporting device read noise. These are statistically distinct quantities with very different meanings, and an attempt has been made to keep these quantities distinct. Using the present readout electronics, near theoretical behavior in error reduction through the use of rereads was observed. These measurements were made at low frequency, which is probably responsible for their success. Prior investigations into error reduction using this method probably employed a sampling scheme such that all noise sources did not appear white. Under these conditions, the theoretically predicted improvement in error will not be obtained.

Fixed pattern noise of the CID 38SG and the CID 17PPRA was examined. Neither device was previously characterized for pattern noise. In addition, a method has been presented for isolating the different sources of pattern noise through the construction of a model of the overall system variance. The second-generation pre-amp-per-row device (38SG) showed an improvement of nearly a factor of two in device pattern noise over the first-generation (17PPRA) device.

The linearity of the photometric response function of the CID 17PPRA was re-examined. The data collected for the studies presented here, as well as data collected by True,[20] Pilon,[10] Bilhorn,[16] and Sims[15], suggest that this result is not clearly attributable to the presence of the on-chip pre-amps, row collection, or decreasing row capacitance, as has previously been suggested. The effect is of considerable concern, as the magnitude of non-linearity depends on the exposure conditions of the device. Hanley[23] looked at both the historical data and a wide variety of newer devices and came to the conclusion that the origin of the linearity problems of the CID camera systems examined in these laboratories over the last 15 years should be considered an open question.

The quantum efficiency of the four devices has been measured. Data taken on the CID 17PPRA indicate that previously reported values are high. Adding a UV down converter to the surface of the device results in improved UV performance while slightly lowering QE in the visible. Data are now available that describe the dark current behavior over the range from 0 to -100°C for the CID 38Q-B and the CID 38SG.

A form of hysteresis in CIDs has been reported and an explanation proposed. A number of other laboratories have now indicated similar problems with the devices. This type of hysteresis is correctable using appropriate bias images after image injection. Charge trap hysteresis has been observed in the more recent generation of devices.

CIDs have previously been found to be uniquely suited to certain specialized forms of scientific imaging, such as atomic emission spectroscopy. In this application, small regions of the device are exposed to light and have excellent performance. Rigorous evaluations of the devices for other types of imaging are unavailable. The data presented here represents a first step toward evaluating CIDs for general imaging purposes.

References

1. G. Michon, *US Pat.*, 3786263, 1974.
2. G. Sims, in *Charge-Transfer Devices in Spectroscopy*, ed. J. V. Sweedler, K. L. Ratzlaff, and M. B. Denton, VCH Publishers, New York, 1994, pp. 1-7.
3. R. S. Aikins, C. R. Lynds, and R. E. Nelson, in *Low Light Level Devices for Science and Technology*, ed. C. Freeman, SPIE Proceedings, Vol. 78, 1976, pp. 65-72.
4. G. R. Sims and M. B. Denton, in *Multichannel Image Detectors*, ed. Y. Talmi, ACS Symposium Series No. 236, Vol. II, American Chemical Society, Washington, D.C., 1983, pp.117-132.
5. J. V. Sweedler, R. B. Bilhorn, G. R. Sims, and M. B. Denton, *Anal. Chem.*, 1988, 60, 282A.
6. J. V. Sweedler, M. B. Denton, G. R. Sims, and R. C Aikins, *Opt. Eng.*, 1987, 26, 1020.
7. J. Zarnowski, J. Carbone, R. Carta, and M. Pace, *Radiation tolerant CID imager*, SPIE Proceedings, Vol. 2172, SPIE, Bellingham, Washington, 1994.
8. G. J. Michon, H. K. Burke, T. L. Vogelsong, and P. A. Merola, *Charge-injection device (CID) Hadamard focal plane processor for image bandwidth compression*, Technical Information Series 77CRD222, General Electric Company, Schenectady, New York, 1977.
9. R. B. Bilhorn and M. B. Denton, *Appl. Spectrosc.*, 1989, 43, 1.
10. M. J. Pilon, PhD Dissertation, University of Arizona, 1991.
11. S. Van Gordon, and J. Carbone, personal communication.

12. Q. S. Hanley, M. B. Denton, E. Jourdain, J. F. Hochedez, and P. Dhez, in *Recent Advances in Scientific Optical Imaging*, ed. M. B. Denton, R. E. Fields, and Q. S. Hanley, Royal Society of Chemistry, 1996, pp. 203-209.

13. L. Mortara and A. Fowler, *Evaluations of Charge-Coupled Device Performance for Astronomical Use.* SPIE Proceedings, Vol. 290, Solid State Imagers for Astronomy, SPIE, Bellingham, Washington, 1981, pp.28-33.

14. J. R. Janesick, K. P. Klaasen, and T. Elliott, *Opt. Eng.*, 1987, 26, 972.

15. G. Sims, PhD Dissertation, University of Arizona, 1989.

16. R. B. Bilhorn, PhD Dissertation, University of Arizona, 1987.

17. G. Sims, in *Charge Transfer Devices in Spectroscopy.* ed. J. V. Sweedler, K. L. Ratzlaff, and M. B. Denton, VCH Publishers, New York, 1994, pp. 9-58.

18. J. Carbone, personal communication.

19. G. Sims and M. B. Denton, *Opt. Eng.*, 1987, 26, 1008.

20. J. B. True, PhD Dissertation, University of Arizona, 1996.

21. J. Carbone, personal communication.

22. G. Sims and M. B. Denton, *Opt. Eng.*, 1987, 26, 999.

23. Q. S. Hanley, PhD Dissertation, University of Arizona, 1997.

24. T. W. Barnard, M. I. Crockett, J. C. Ivaldi, P. L. Lundberg, D. A. Yates, P. A. Levine, and D. J. Sauer. *Anal. Chem.*, 1993, 65, 1231.

25. S. Strunk, P. Chen, M. Fattahi, T. Kaysser, B. Nguyen, H. Tseng, R. Winzenread, and M. Wei, *A 4 Million Pixel CCD Array for Scientific Applications*, SPIE Proceedings Vol. 1900, Charge-Coupled Devices and Solid State Optical Sensors III, SPIE, Bellingham, Washington, 1993.

26. M. Joy, personal communication.

27. T. Jovin, personal communication.

DEVELOPMENT OF A BACK-ILLUMINATED 4096 × 4096 15-MICRON PIXEL SCIENTIFIC CCD

Michael Lesser
Steward Observatory
University of Arizona
Tucson, AZ 85721

Richard Bredthauer
Semiconductor Technology Associates[*]
32681 Rachel Circle
Dana Point, CA 92629

1 INTRODUCTION

The National Science Foundation has funded development of an 8k × 8k CCD camera, prime focus optical corrector, and filter and guider system to be placed at the University of Arizona 2.3-m telescope on Kitt Peak (near Tucson, Arizona). We describe in this paper the CCD design and characterization effort that has lead to the selection of CCD detectors for this system.

A schematic view of the camera's optical design is shown in Figure 1. The one-degree field (180 mm diagonal) will produce 0.47 arc-seconds/pixel using 15 μm pixels in the f/2.7 beam. A filter change mechanism will be placed between the last optical element and the detector dewar window. We will have a selection of six 6 × 6 in filters. The guide system to control telescope tracking will use a pick-off mirror located above the filter plane to direct unfiltered light to a frame transfer CCD running at a frame rate of approximately 1 Hz. Centroiding of star images in real-time will provide feedback to the telescope control system to track long integration time exposures.

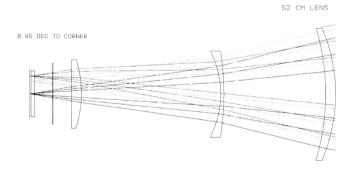

Figure 1 *The optical layout of the prime focus camera. The CCD focal plane is at the far left, followed by the dewar window, filter wheel, and three optical elements*

---

[*] previously with Lockheed Martin Fairchild Systems

2 4096 × 4096 CHARGE-COUPLED DEVICES

The current astronomical standard for large-format CCDs is 2k × 4k 15 μm pixel devices, such as those that are now or have been produced by SITe,[1] EEV,[2] Lincoln Labs,[3] Orbit,[4] and Loral.[5]

Lockheed Martin Fairchild Systems (LMFS, Milpitas, CA) has produced the commercial-grade CCD485 4k × 4k 15-μm pixel CCD for several years. These devices have been made with good yield and have excellent cosmetic quality. They are not, however, optimized for backside processing or low noise readout. Our initial interest in these devices has been to determine if we could use 4k × 4k CCDs for our camera instead of the more common 2k × 4k devices.

In our experience with astronomical CCD foundry runs, yield is not closely related to device size for single lot runs. More often, poor yield is due to mask errors, design errors, poor silicon quality, or other lot-wide problems that lead to excessive shorts or other failures. While we recognize that for large-volume production there is a much more direct inverse relationship between die size and yield, we believe that for this project a factor of two increase in size is not of great yield concern.

There are several advantages for using as large a device as possible in a mosaic camera. Having fewer devices requires less calibration, backside processing, and handling; generally simplifies I/O wiring; and decreases overall system complexity. We are also interested in pushing scientific imagers to larger formats for future projects. Because of these factors and the high quality of the commercial CCD485 part, we have chosen an optimized 4k × 4k CCD as our baseline detector for low light level imaging. The camera's final focal plane will consist of a 2 × 2 mosaic of such devices.

2.1 CCD485 Characterization

Because of fixed funding, we must be fairly certain we can obtain scientific grade devices of this large format for this project. We therefore have characterized a CCD485 device to obtain a baseline performance specification for the scientifically optimized devices. Several other groups have also used CCD485 devices in scientific applications and have informally reported similar characteristics. Results of our characterization are presented here.

The device tested was packaged in the commercial ceramic carrier by LMFS. It was mounted in a general purpose test dewar used for characterizing large format devices at the Steward Observatory CCD Laboratory. Tests were performed at –90 C. Measurements were made using a San Diego State University (SDSU) Generation I CCD controller read out through one amplifier. The pixel rate was approximately 50 kHz.

Charge-transfer efficiency was measured to be 0.999992 (parallel) and 0.999994 (serial) using an Fe-55 X-ray source (1620 electrons). A parallel CTE histogram plot is shown in Figure 2. It is clear that the silicon quality used to fabricate the CCD485 devices is acceptable for scientific use. Additional CTE tests were performed at –150 C to evaluate the silicon quality. Very little degradation was measured.

Read noise was measured to be 7.5 electrons, calibrated with an Fe-55 source. Full well capacity was 70,000 electrons in MPP mode. Additional optimization of clock voltages can be expected to further improve these values. The cosmetic quality of the device tested was excellent, even though it was not in LMFS's highest-grade category. We found only four partially blocked columns due to pixel traps. Overall the device

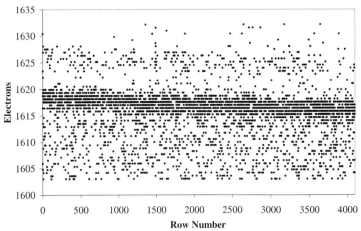

Figure 2 *Parallel CTE histogram plot of an Fe-55 X-ray image taken with the CCD485 4 × 4k CCD. The slope of the line indicated the CTE value (0.999992).*

tested was of excellent quality and demonstrated that a similar but "scientifically" optimized 4k × 4k would perform very well for the 8k × 8k camera.

3 OPTIMIZED DETECTOR REQUIREMENTS

In this section we describe the requirements we have placed on the scientific version of the 4k × 4k CCD. These requirements have been incorporated into a new design, and at least one lot run will be fabricated this year.

We require that the entire mosaic readout time be less than 60 s, with read noise under 5 electrons rms. We will use an SDSU Generation II CCD controller.[6] If the final device pixel rate must be as low as 50 kHz to achieve good noise performance, we will need four amplifiers per device. We hope to run faster and read out the mosaic through two amplifiers per CCD, but we are specifying a requirement of four amps per CCD.

We require high U-band quantum efficiency (QE) as well as high overall (broadband) QE, necessitating backside-illuminated devices. We will initially backside process the devices in our Steward Observatory CCD Laboratory using techniques already developed.[7,8,9,10,11]

We have decided against requiring the devices to be "edge buttable" in the traditional sense. Most edge-buttable designs have only two amplifiers (at the ends of one serial register). This reduces readout time and the availability of good amplifiers in the case of low amplifier yield. Buttable devices often have reduced parallel clock buss size and/or number, which can limit speed and versatility (e.g. no separate frame transfer region). There are several post-fabrication yield issues with buttable CCDs that we would like to avoid, including increased dicing damage (chipping), difficult handling, and problems associated with die level thinning very close to the device edge.

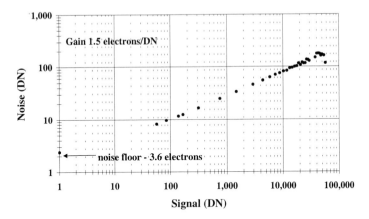

Figure 3 *Measured photon transfer curve of a candidate amplifier for the scientific 4k × 4k device. Subsequent tests showed a read noise floor of just under 3.0 electrons rms*

In astronomical imaging, multiple images, often taken for cosmic ray rejection, act to average the non-imaging gaps. We have therefore decided to minimize the dead (inactive) edge but do nothing to decrease *overall* device yield. This decision has led to a design that has two serial registers opposite each other and a 500-μm-wide inactive region from the first column to the edge of the device on the two parallel sides. The inactive width on the two serial register sides is 1000 μm.

3.1 Amplifier Selection

Previous generations of Loral scientific imaging devices had minimum noise of ≈6 electrons. While direct imaging does not require extremely low noise due to sky background, we would like to allow these devices to be used for other spectroscopic applications without redesign. We also need to read the devices as fast as possible for efficient use of telescope time. These two specifications require a new amplifier to be used for the scientific imaging version of the 4k × 4k CCD.

Several test devices with new low noise amplifiers were manufactured in recent years and have been tested as candidates for the new device. The chosen design has a measured noise of 3 electrons rms, calibrated with Fe-55 X-ray imagers. A photon transfer curve of the 64 × 64 pixel test device is shown in Figure 3, taken at –120 C and a read rate of 50 kHz.

3.2 Epitaxial Thickness and Resistivity

QE and resolution are affected by both epitaxial thickness and resistivity. The critical parameter, which can be controlled after device fabrication, is the final device thickness after thinning, not the epitaxial thickness itself. Any undepleted field-free region between the frontside depletion edge and back surface significantly degrades the modulation transfer function (MTF) due to charge diffusion to adjacent pixels. Devices should be thinned to near the depletion edge to avoid this field-free region. Unfortunately, devices cannot be arbitrarily thinned into the epitaxial layer, because thinning defects increase with additional etching. We have found our thinning yield drops after etching more than a few microns beyond the epitaxial-substrate interface. For this reason, the starting material epitaxial thickness should be no more than a few microns greater than the desired final device thickness. In addition, the resistivity of the epitaxial layer should be such than nearly all of the device is depleted (depletion depth increased as the square root of resistance). These requirements set the starting material specification. We have specified 10-μm-thick epitaxial material of at least 100 Ohm-cm resistivity.

3.3 Final Device Specifications

Below is a list of the final design parameters for the 4k × 4k devices we will have fabricated for this project.

- 15-μm-square pixels
- four low noise (< 3 electrons rms) amplifiers, one in each corner
- split serials and parallels for 1, 2, or 4 readouts per device
- full well capacity of at least 100,000 electrons
- front side ground contact for back-illuminated processing
- 500-μm image pixel to device edge distance (minimum) for die thinning
- MPP support
- 10-μm epitaxial thickness, > 100 ohm-cm silicon

4 BACKSIDE PROCESSING

Initial thinning will be performed at the Steward Observatory CCD Laboratory using our standard backside process. We have already developed the thinning techniques for single die as large as 70 mm per side. Similar-sized devices we have processed include the Caltech BIG CIT 4096JJ 4k × 4k CCD and the ESO VLT 5 CCD hybrid device, both of which we have already successfully thinned and operated.

A silicon bonding support will be used for expansion matching of the detector, an important requirement for cryogenic operation of such large area devices. The custom polished substrates also allow a flat imaging surface to be obtained after thinning.

4.1 Backside Charging

We will utilize the Arizona chemisorption-charging technique developed for maximum QE and QE stability. We have previously applied this coating to many large area devices. The process is based on the chemisorption bonding of negative ions to thin metal films.[12] We show in Figure 4 the QE of several chemisorption-charged devices. This charging mechanism is stable against any environment conditions we have found and seems an excellent choice for the 4k × 4k CCDs.

Chemisorption Charged CCDs
Various Manufacturers

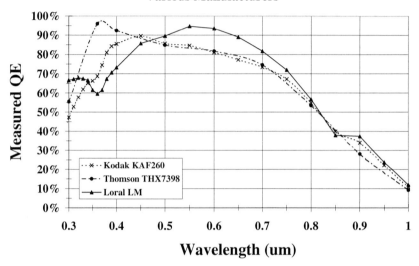

Figure 4 *The measured QE of several CCDs thinned in the Steward Observatory CCD Laboratory. The differences in the curves are due to the different antireflection coatings applied. These are typical values expected for the final thinned 4k × 4k devices*

5 PACKAGING

We will utilize packaging technology already developed in the Steward Observatory CCD Laboratory for other very large area devices. These devices have included the Philips 7k × 9k 12-μm (140 mm diagonal), BIG CIT 4k × 4k 15-μm (90 mm), and ESO VLT (5 CCDs processed together—80 mm) CCDs. Our basic packaging system includes an Invar-36 package, an FR-4 printed circuit fan out, and a 37-pin Nanonics connector.

Figure 5 shows a schematic layout of the 8k × 8k mosaic of four devices. The gap between active imaging areas is 6 mm. The rows and columns of adjacent devices are rotated 90 degrees relative to each other in order to minimize this gap.

The Invar carrier is machined and polished flat to a specification of 2.5 μm peak-to-valley. The flatness goal of the entire mosaic is 20 μm peak–valley. We have measured similar Invar packages to have less than 1-μm peak–valley flatness deviation. The packages and attachments have been successfully rapidly thermally cycled between liquid nitrogen (77 K) and room temperature. We have imaged large format devices packaged in this manner as cold as -165 C. The CCDs will be attached to the Invar carrier with thermally conductive thermoplastic.

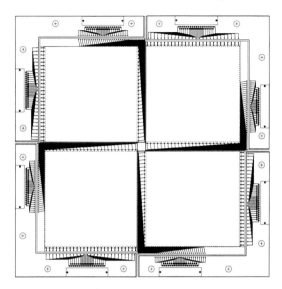

Figure 5 *A schematic layout of the 8k × 8k prime focus CCD mosaic. Note the CCDs are rotated relative to each other to reduce the inter-device gap. All four amplifiers on each CCD are used, through two connectors per package. The four packages are mounted individually to a single cold plate*

Acknowledgements

Support for this project is provided by NSF Grant AST-9871490. The focal plane packaging was designed and developed by David Ouellette of the Steward Observatory CCD Laboratory. James Burge (University of Arizona) developed the optical design. We thank Ed Olszewski and Gary Schmidt for useful discussions and all members of the CCD Laboratory for their assistance in developing the backside processing techniques described.

References

1. T. W. Woody et al., *Proc. SPIE*, 1997, **3019**, 189.
2. P. S. Heyes, P. J. Pool and R. Holtom, *Proc. SPIE*, 1997, **3019**, 201.
3. M. Wei and R. J. Stover, *Proc. SPIE*, 1998, **3355**, 598.
4. R. J. Stover, W. E. Brown, D. K. Gilmore, and M. Z. Wei, *Proc. SPIE*, 1994, **2198**, 803.
5. G. A. Luppino, R. A. Bredthauer, and J. C. Geary, *Proc. SPIE*, 1994, **2198**, 810.
6. 6. R. W. Leach, F. L. Beale, Sr., and J. E. Eriksen, *Proc. SPIE*, 1998, **3355**, 512.
7. M. P. Lesser, *Proc. SPIE*, 1991, **1447**, 177.
8. M. P. Lesser, A. Bauer, L. Ulrickson, and D. Ouellette, *Proc. SPIE*, 1992, **1656**, 508.
9. M. P. Lesser, *Proc. SPIE*, 1993, **1900**, 219.
10. M. P. Lesser, *Proc. SPIE*, 1994, **2198**, 782.
11. M. P. Lesser and D. B. Ouellette, *Proc. SPIE*, 1995, **2415**, 182.
12. M. P. Lesser and V. Iyer, *Proc. SPIE*, 1998, **3355**, 23.

ROLIS: A SMALL SCIENTIFIC CAMERA SYSTEM FOR THE ROSETTA LANDER

H. Michaelis, T. Behnke, M. Tschentscher, H.-G. Grothues and S. Mottola

Institute of Space Sensor Technology and Planetary Exploration
DLR–German Aerospace Center
Rudower Chaussee 5
D-12489 Berlin
Germany

1 INTRODUCTION

The camera group of the DLR Institute of Space Sensor Technology and Planetary Exploration is developing specific imaging instruments for scientific space missions, ground based astronomy and other space and airborne applications. The key subjects of the group are as follows:

- Investigation in new imaging technologies (sensor configurations, imaging concepts, evaluation and specification of new imaging detector arrays)

- Conception, design and development of imaging systems for space and ground-based application

- Investigation and design of high performance sensor electronics (e.g. miniaturisation, harsh environment, low noise, high data rates).

   The members of the camera group have an average of over 15 years experience in the design and development of imaging instruments. One example is the development of sensor electronics for the Cassini mission (DISR) and the imager for Mars Pathfinder (IMP), which was developed in Germany at the Max-Planck-Institute for Aeronomy (led by Uwe Keller). Another example is the development of the High Resolution Stereo Camera (HRSC) at the DLR (led by Gerhard Neukum), which was originally designed for the Mars-96 mission.
   One of the present main activities of the DLR camera group is the participation in the Rosetta mission with the VIRTIS experiment (an imaging spectrometer) and the imaging experiment ROLIS.

2 THE ROSETTA MISSION

ROSETTA is an ESA cornerstone mission that will be launched in the year 2003 on an Ariane 5. After a long cruise phase, the satellite will rendezvous with comet Wirtanen

Figure 1 *CCD detectors used in DLR-imaging instruments*

Figure 2 *The Rosetta spacecraft*

and orbit it, while taking scientific measurements. A Surface Science Package (SSP) will land on the comet's surface to make in-situ measurements. During the cruise phase, the satellite will be given gravity assist manoeuvres, once by Mars and twice by the Earth. The satellite will also make flyby measurements of two asteroids.

Figure 3 *Nucleus of comet Halley taken by Giotto-HMC (courtesy of MPAE, U. Keller)*

Figure 4 *The Rosetta Lander (STM during integration)*

3 THE ROSETTA LANDER

After a landing site is chosen, in August 2012 the Rosetta Lander, RoLand for short, will land on the comet's surface and the Surface Science Package (SSP) will be released to perform in-situ measurements.

RoLand will land with a relative velocity of less than 1 meter/second and will then transmit data from the surface to the orbiting spacecraft, which will relay it to Earth. A successful Rosetta Lander mission will allow operation of the major instruments for a time sufficient to analyse the time dependence of chemical and physical characteristics at the surface of the nucleus. Substrates for chemical analyses will be taken from the surface and from different depth down to at least 200 mm. Sufficient time will be available to study the time dependence of characteristics like temperature, thermal conductivity, electrical conductivity, and so forth. Medium duration measurements will be performed, such as acoustic and seismic investigations and nucleus sounding during several orbits of the Rosetta spacecraft.

Eight instruments will be placed on the lander to investigate the elemental, isotopic, molecular and mineralogical composition and the morphology of early solar system material as it is preserved in the cometary nucleus.

The imaging system of the Rosetta Lander is called CIVA/ROLIS. In the following section, one aspect of this system will be described in greater detail to demonstrate the new technical approach and its potential application to other imaging systems.

4 ROLIS

One of the core payload instruments of the Rosetta Lander is the ROLIS imaging system, which consists of the Descent- and Downlooking Imager (ROLIS-D), the Imaging Main Electronics (ROLIS/CIVA-IME) and an optional Close-Up-Imager (ROLIS-CUI).

The measurement objectives of ROLIS are as follows:

- Characterisation of surface features on scales from meters to fraction of millimetres (particle size distribution, cracks, pores, vents, mocrocraters, and so forth)
- Identification of compositional and textural small-scale inhomogeneities through colour imaging
- Monitoring of the diurnal and long-term changes in the cometary surface
- Support of the selection of drilling areas
- Giving context to the measurements of the in-situ analysers
- Bridging the gap between the orbiter and the microscope data

The Descent- and Downlooking Imager (ROLIS-D) will contribute to understanding of the cometary regolith by imaging the landing area with a resolution ranging from 20 mm/pixel to 1 mm/pixel during the descent phase and multispectral 3-D close-up imaging of selected areas below the lander with resolutions of 0.25 mm mm/pixel.

Figure 5 *The cometary surface—an artists image-painting by William K. Hartmann*

Table 1 *Technical Requirements of ROLIS-D*

Item	Spec
Resolution (IFOV)	1 mrad/pix
Field of View (FOV)	1 rad (60° x 60°)
Focus	adjustable (infinity, close up)
Depth of Focus	≈ +/- 100 mm @ 300 mm
Multispectral Imaging	(≥4): red, green, blue, NIR
High dynamic range	> 80 dB
Good SNR	100 (low noise)
Low mass	400g (incl. electronics)
Low power	< 2 W
Low temperature operation	≈130°K (100°K)
Simple el. interface, low thermal conductivity	no separation between detector and detector electronics
Protection	against radiation, dust, condensates

4.1 The Technical Concept of the ROLIS-D Camera

In the following, we briefly describe the technical requirements, sensor electronics, optomechanics and data processing unit of ROLIS.

4.1.1 Technical Requirements. The ROLIS imaging system has to withstand harsh environmental conditions without major performance degradation. One of the main requirements of ROLIS-D is to survive and operate at temperatures down to 100°K–150°K with a minimum of power consumption (2–3 W). The optics must provide high-resolution images at ranges between infinity (during descent) and about 0.3 m (after landing), with a field of view of about 60° and a resolution of 1 mrad (1mm/pixel at 1 m distance). After landing, multispectral images have to be taken in at least 4 spectral channels (red, green, blue, NIR). Furthermore, the instrument has to survive the long cruise to comet Wirtanen (8 years) and the cosmic radiation, at a mass budget of only 400 grams. The main technical requirements are summarised in Table 1.

4.1.2 The Modular Sensor Electronics of ROLIS. In order to fulfil the technical requirements, a specific concept was developed, the Modular Sensor Electronics System, referred to by its acronym, MOSES.

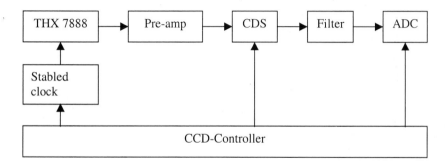

Figure 6a *Block diagram of the CCD sensor electronics (MOSES-THX 7888)*

Figure 6b *Breadboard of MOSES-THX 7888*

The technical concept is based on a rigid-flex 3D-interconnection between the functional electronic boards (CCD head, focal plane assembly, clock driver, signal chain, and interface). The technical solution allows the operation of the complete electronics at temperatures down to 130K. The baseline is the FT 1K × 1K CCD THX7888 manufactured by THOMSON. Figure 6 shows a block diagram of the CCD sensor electronics.

In order to measure the electronics noise without the CCD, a technique is used in which the rms noise of an A/D converter (including all components), referred to the input due to histogram evaluation at fixed input signal, is determined. An important assumption is that the model for the noise generation is a Gaussian distribution source added to the input of the electronics. The probability density function (PDF) of an input shows that the majority of the output codes will occur in a single bin, but that there must be additional codes corresponding to the tails of the distribution. The fraction that occurs outside the main code depends on the spread, or standard deviation σ, of the noise distribution. For a symmetrical Gaussian PDF, the equation

$$P = \frac{2}{\sigma\sqrt{2\pi}} \int_{-\infty}^{-0.5LSB} e^{-x^2/2\sigma^2} dx \tag{1}$$

expresses the probability of getting a code outside the main bin for that PDF. The terms inside the integral are simply the expression for a Gaussian distribution with unit area under the curve and standard deviation, σ. We integrate from -∞ to −0.5LSB to find the area under one tail, and from +0.5LSB to +∞ for the other. By symmetry, one can simply multiply the area under one tail by 2. To ultimately determine the noise equation (5) must be solved for σ with P determined from the histogram as the fractional portion of code hits outside the main bin. Unfortunately, the integral has no closed form solution in terms of σ. One option is to iterate σ and numerically evaluate equation (5) until the solution equals the fraction determined from the histogram. Alternatively, we can fix σ, iterate the upper limit of integration, plot the result and use it to find the upper limit that corresponds to any value of P. Because we already know the value of P from the histogram, we can simply find the x-axis point, x_0, corresponding to P. This value is

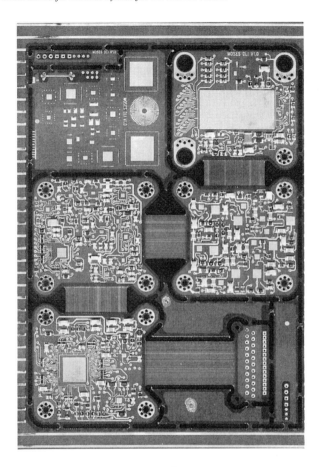

Figure 7 *The MOSES 3-D detector electronics in unfolded configuration*

Table 2 *Parameters of the THX7888 1Kx1K FT CCD and readout*

Parameters	Value	Remarks
Pixel readout time	3.2 μs	
Resolution	14 bit	
Total noise in darkness	13 e⁻	@200°K
CCD noise	12 e⁻	@200°K
Electronic noise	5 e⁻	
Responsivity	7 μV/e⁻	
System gain	20 e⁻/DN	
Dark current	1700 e⁻/sec 1100 e⁻/sec 2000 e⁻	Image section @ 10°C Memory section @10°C Serial register
Antiblooming control	Yes	
Peak QE	18%	

equivalent to having evaluated equation (5) from $-\infty$ to $-0.5LSB$, we had known the correct value for σ.

To calculate the rms noise, we simply solve the equation

$$-x_0\sigma = 0.5LSB \tag{2}$$

For example, if we have a measurement with n samples (n should be large) and k code hits outside the main bin, the value k/n represents

$$F = \frac{2}{\sigma\sqrt{2\pi}} \int\limits_{-\infty}^{x} e^{-z^2/2\sigma^2} dz. \tag{3}$$

Using a standard normal curve, or z-Table, we find x_0 and calculate σ (equation 3).

With the application of limited resources concerning mass, volume, power, and the requirement of high reliability of components for space instrumentation, we obtain the key parameters of the CCD-camera, identified in Table 2.

4.1.3 Features of the Optomechanics of ROLIS-D. A special feature of the concept is the bimodal optics, which enable good focusing, with only a simple mechanism, during the descent and after landing. During the descent phase, a so-called infinity lens is in front of the fix-focus optics of the CCD. It acts like an eyeglass and shifts the focus of the fix-focus-optics to infinity. Furthermore it operates like a dust cover and saves the main optics from dust contamination and condensed material during the descent phase and during the non-operation periods on ground. During the operation phases after landing, the infinity lens will be turned out of focus of the main lens by a special excenter mechanism driven by a low temperature qualified stepper motor (see Figure 8).

In order to provide multicolour images with good signal-to-noise ratio (SNR>100) of the dark cometary surface, a highly integrated, low power LED illumination is used with up to 64 LED chips per colour. The high integration density is achieved by chip-on-board technology similar to the MOSES detector electronics.

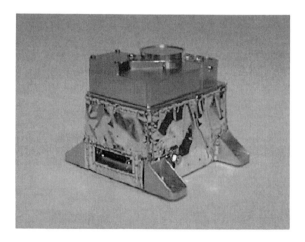

Figure 8 *The ROLIS-D STM unit with infinity lens closed*

Figure 9 *The miniaturized flat field illumination of ROLIS-D with LED arrays*

The main characteristics of the ROLIS-D camera are summarised as follows:

ROLIS Camera Head

- Detector: Thomson 7888A CCD, 1024 × 1024 14-µm pixel
- Frame transfer architecture. Exposures 1 ms–52 s
- A/D conversion: 14 bits.
- Bimodal optics: infinity mode/close-up

Focal length	f#	FOV	working dist.	working dist.
12 mm	5.0	~1 rad	>1.25 m	300±50 mm

- Motor-actuated infinity lens.
- Illumination device: 8 tungsten lamps with RGB filters and quasi-parabolic reflector
- Mass 350 g
- Average power 1.6 W.

4.1.4 The ROLIS/CIVA Imaging Main Electronics. All imaging subunits of ROLIS and CIVA are controlled by the Imaging Main Electronics (IME). The IME is located inside the Rosetta Lander and has serial command/data links to all camera heads. The IME consists mainly of a processor/control board—the so-called Common Data Processing Unit (CDPU) and a 16 Mbytes frame buffer, called mass memory. The CDPU is based on the radiation hard RTX2010 processor and is programmed in Forth. During the descent phase, the data of ROLIS-D will be continuously stored in the mass memory as in a ring-buffer: when a new image is taken in, the last one in the image buffer will be removed. The mass memory is based on highly integrated SRAM cubes with a capacity of 4 Mbytes per cube. Before transmission of the image data to the Rosetta spacecraft data compression will be performed. The data compression is based on a wavelet transformation and enables a compression ration of about 10 with good image quality. The whole ROLIS-IME weights only 570 grams and has an average power consumption of 1.4 W.

5 OTHER APPLICATIONS OF ROLIS SENSOR ELECTRONICS

A big advantage of the MOSES sensor electronics of the ROLIS imager is the modular architecture. It enables high flexibility and simplifies the adaptation to other CCD detectors, and other specifications, such as implementation of special readout modes (e.g. TDI) and different readout rates, in a very cost-effective way. Therefore, it is possible to use MOSES, without change in hardware configuration in the high performance camera system, for automatic fire detection in forests (AWFS). This application is a real spin-off effect of the ROLIS development and is operating very successfully during the experimental verification phase in the forests of Berlin-Brandenburg (Germany).

Another example is the application of MOSES in the GEMINI imaging system for the ESA technology mission, SMART-1. In this application, it was proposed to implement MOSES in a wide-angle camera and in a high-resolution imager. In the wide-angle camera, MOSES is operating with slow readout rates of about 400 k sample/sec and low noise 12 e rms (with the Thomson CCD THX7888). The high-resolution imager operates

Figure 10   *The Common Data Processing Board of ROLIS with Harris-RT-2010 processor (100mm × 120mm)*

Figure 11 *The camera for automatic fire detection in forests with MOSES-THX7888 electronics*

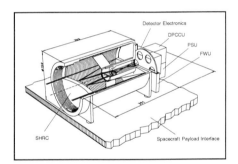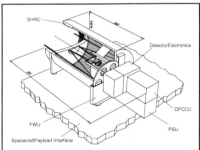

Figure 12 *SMART-1: GEMINI Imaging System*

with 4 or 8 M sample/sec and has the possibility to run in TDI mode to give a high signal-to-noise-ratio (SNR) even at very low dwell times. This is especially important during the originally planned asteroid fly-by of the SMART-1 mission. It is also intended to fly the high-resolution image for GEMINI on the European Mars-Express Mission in 2003 (see Figure 12).

The DLR camera group is also developing CCD imaging systems for telescope observations of asteroids and especially near-Earth objects (NEOs). In order to achieve the required large field of views, the CCD detector area has to be as large as possible. The largest CCD detector investigated and used at the DLR is the Philips 7k × 9k CCD, as shown in Figure 13. Another approach is CCD mosaics as typically used in state-of-the-art astronomical CCD cameras. For high readout rate and low noise, it is advantageous to implement the sensor electronics rather close to the CCD detector. Therefore, it was proposed to implement small modules of the (low temperature

Figure 13 *Philips 7kx9k CCD with high integrated detector electronics on the focal plane*

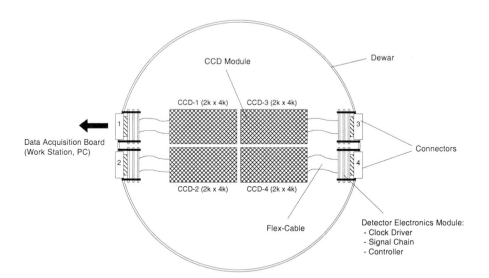

Figure 14 *Schematic view of the proposed concept for CCD mosaic cameras*

qualified) MOSES sensor electronics inside the CCD dewar. Because of the small dimensions of MOSES, it could be easily assembled at the electrical connectors for each CCD detector. This concept would also simplify the interface and wiring for each CCD of the mosaic, because only a serial digital interface has to be handled for the image data transmission and CCD control (see Figure 14).

6 SUMMARY

The idea of this presentation was to show some of the activities running presently at the DLR Institute of Sensor Technology and Planetary Exploration for the Rosetta Mission. Features of the technical concept of the ROLIS- Descent- and Downlooking Imager, and their potential applications in other imaging systems, were described.

Acknowledgements

We gratefully acknowledge the Deutsche Agentur für Raumfahrtangelegenheiten (DARA) and the Bundesministerium für Bildung, Wissenschaft, Forschung und Technologie for financial support of this work.

Reference

1.   H. Michaelis, H.-G. Grothues, T. Behnke, et al., ESA SP-423, 1998.

DETECTION OF ISOLATED MANMADE OBJECTS USING HYPERSPECTRAL IMAGERY

Walter F. Kailey and Michael Bennett

Ball Aerospace & Technologies Corp.
1600 Commerce St.
Boulder, Colorado 80020 USA

1 ABSTRACT

An algorithm for automated detection of isolated manmade objects using hyperspectral images gathered from airborne or spaceborne platforms is discussed. This needle-in-a-haystack problem is of interest to many government agencies throughout the world, as it arises in law enforcement, control of borders, counter-drug enforcement, and military surveillance applications. Our algorithm is based on linear least squares modeling. Algorithm performance in detecting marginally unresolved targets not recognizable to a human observer has been measured using HYDICE imagery. Signal-to-noise and spatial resolution requirements for detection under various circumstances are discussed. These results allow the tradeoff between spatial and spectral resolution for surveillance sensors to be quantitatively evaluated, perhaps for the first time. Performance of our least squares algorithm is compared to a closely related orthogonal projection method and found to be superior.

2 INTRODUCTION

This paper discusses the use of hyperspectral imagery to solve what might be termed the "needle in a haystack" problem. Small objects, often man-made objects in remote areas, different from the background scene, may be present in very low density over a very large area. The problem is to develop an automated method for detecting them that does not require a human image interpreter to study the very large numbers of high-resolution images needed to cover the entire search area. This application is sometimes referred to as automated target detection (ATD), which is distinguished from automated target recognition (ATR) by the level of classification involved. In general, ATR classifies targets much more precisely than ATD.

ATD algorithms may be useful for search and rescue applications, counter-drug applications and border control generally, law enforcement, and support to military operations. ATD can be used to cue high resolution imaging sensors and can then be followed up with ATR and/or human image interpretation.

3 DATA COLLECTION

Hyperspectral images are collections of images of the same scene in a large number of spectral bands. Such images are collected using an imaging spectrograph. Typically, this type of instrument images the scene onto a narrow slit, recollimates the image of the slit, spectrally disperses it in the direction orthogonal to the slit, and then re-images it onto a two dimensional focal plane array. The result is an image with spectral information in the cross-slit direction and spatial information in the other direction. When the slit is scanned across the scene, *e.g.* from an airborne platform, and a series of such images is collected, a hyperspectral image or *data cube* is obtained.

Other ways of obtaining data cubes include the use of an imaging Fourier transform spectrograph (FTS), Descour's rotating prism spectrograph,[1] and linearly variable filters. With the exception of linearly variable filters, all of these alternative techniques have the "multiplex advantage", as well as the "multiplex disadvantage". They also suffer from the disadvantage that consecutive images read out from the focal plane must have an accurately controlled spatial registration in order to create a true data cube.

The multiplex advantage is defined as the advantage that comes from maximizing the simultaneous collection of photons over a two-dimensional field of view and over the whole spectral range, *i.e.* over all three dimensions of the data cube simultaneously. Since the collection device is two dimensional, the way in which the three dimensions of data are projected onto the focal plane must be systematically varied over a series of images so that the three independent dimensions can be recovered from the data set by means of an algorithm.

It is rarely stated, but true nonetheless, that the multiplex advantage is only helpful when the instrument performance is detector system noise limited, at least in part. To realize this intuitively, think about an imaging FTS, as compared to a linearly variable filter or dispersive slit instrument. The latter two instruments view only one wavelength channel at a time and therefore produce a lower irradiance level on the detector. Thus, it is easier for the FTS, viewing a broad band source, to overwhelm the detector system's read noise with photon noise. The additional photons, however, increase the noise spectral density by the square root of the number of spectral channels, assuming a flat spectrum source. This increase in noise is exactly compensated (within the square root of two, anyway) by the fact that every spectral channel is observed multiple times in the series of images that makes up the interferogram. Thus, if a dispersive slit or linearly variable filter spectrograph can attain the background limit of performance, there is no multiplex advantage.

Then, too, if the source spectrum is far from flat, as with ambient temperature terrestrial sources observed in the 3–10 μm region, one has a multiplex disadvantage at the dimmer wavelengths. This disadvantage is defined as the noise averaging across spectral channels that is inherent in multiplexed instruments.

Among multiplexed instruments, the rotating prism has the unique disadvantage of requiring a focal plane that is larger than either the spatial or spectral dimension of the data cube in both of its dimensions. This is because the instrument produces spatial and spectral dispersion in the same dimension within the image. By varying the spectral dispersion direction within a two-dimensional image, a series of images is obtained, related to the data cube by a known linear transformation that can be inverted. Of course, the data cube will not be spatially registered, unless the camera can track the scene to subpixel accuracy over the entire field for at least one-fourth the period of rotation of the prism. Finally, it is not clear how the rotating prism can be used in pushbroom mode, *i.e.*

with continuous spatial scanning, while the step-stare procedure is relatively inefficient and expensive and is not favored for broad-area search.

The imaging FTS, whether used in pushbroom or in step-stare mode, has a similar limitation. To step-stare, image motion must be frozen out by a control system while the complete range of interferometer path differences is sampled. In the pushbroom mode, the image is scanned continuously across one dimension of a two-dimensional focal plane, while the interferometer path difference is continuously varied at the same time. The dimensions of the focal plane are such that a complete two-sided interferogram is obtained in the time needed for any point on the ground to transit the focal plane. This requires accurate alignment of the scanning direction to one focal plane axis and also accurate synchronization of the scanning motion with the interferometer's moving element. Use of a spatially modulated FTS, in which the path difference is a fixed function of field angle at the focal plane, can alleviate the latter constraint.

It is obvious that the image motion control problem is the same for the linearly variable filter as for the spatially modulated FTS, except that the FTS must typically maintain the desired level of registration over more than twice as many samples.

The multiplex advantage is often of great practical value, depending on the application, since detector technology does not always make it possible to readily attain the background noise limit. On the other hand, this advantage is obtained at the expense of better platform attitude knowledge and control requirements. Thus the preferred instrument configuration will vary from one application to another.

The images used for this work were collected by HYDICE,[2] which is a dispersive slit imaging spectrograph on an airborne platform.

4 LINEAR UNMIXING APPLIED TO ATD

The basis of our algorithm is the simple idea that each pixel's spectrum is comprised of a weighted sum of the spectra of objects imaged onto that pixel. This model does not account for the nonlinear affects of multiple scattering, but for objects visible from overhead that are seen in reflectance, it is clearly a good approximation in most instances. This basic linear model can be extended to include infrared thermal radiation, as well, with only a slight increase in complexity.

The spectra of materials that comprise the background scene are referred to as end-members. Thus, the model posits that a background pixel is a linear combination of end-members, while a pixel containing a target is a linear combination of end-members plus the target spectrum. The target spectrum is not assumed to be completely deterministic or completely known. However, obtaining good results in highly cluttered or vegetated scenes does require that a significant portion of the target spectrum satisfy these criteria.

The algorithm we are using has been described mathematically elsewhere.[3,4] It consists of two filters computed in parallel during a single pass over the image. The first filter, designated χ^2, is the traditional measure of statistical likelihood, which is also the squared Euclidean distance between the model and the data. The model, in this case, assumes that the pixel being operated on is a linear combination of end-members derived locally within the image (see below). The data is a vector consisting of the measured spectrum of the pixel being operated on, where each channel has been normalized by an estimate of its standard deviation. This estimate should consider both measurement noise and, when appropriate, inherent stochastic variation of the end-members themselves. The χ^2 filter is a spectral anomaly detector. It is very sensitive to the presence of small, isolated man-made objects in wilderness areas such as oceans, deserts, forests, or grasslands. It is also insensitive to lines of communication, such as roads and telephone lines. Appropriate definition of the local areas from which end-members are drawn also ensures that large man-made structures, such as farmhouses and barns, will not cause false alarms, so it is equally effective in agricultural areas.

The χ^2 filter takes no account of known spectral properties of the target. If these are partially known, performance can be greatly improved by augmenting χ^2 with a second filter, which we term the K filter. The K filter is the conventional statistical measure of the explanatory power of a basis vector used in linear least squares fitting procedures. It is the ratio of the basis vector's coefficient to the square root of its auto-covariance estimate and is sometimes referred to as the statistical significance of the coefficient.

The method of detection we use declares a target whenever:

1. χ^2 exceeds a threshold, *and, when applicable,*
2. K exceeds a threshold in two or more adjacent pixels, one of which is coincident with the χ^2 excess.

The χ^2 filter is usually computed first, as it is of lower computational complexity, requiring only the end-members as basis vectors. The K filter is then computed over local areas adjacent to χ^2 excesses. As with any ATD system, both thresholds are dynamically adjusted to maximize sensitivity within the constraint of available computational throughput and other resources, resulting in a constant false alarm rate.

The K filter is closely related to the Orthogonal Subspace Projection (OSP) operator developed by Harsanyi, *et al.*[5,6] However, there is a subtle computational difference between them, which is unimportant in low clutter scenes. Our most recent results (see below) seem to indicate that K is significantly more sensitive in the presence of vegetative clutter. Moreover, pre-filtering using χ^2 brings further significant improvement over the OSP in highly cluttered scenes. Our numerical experiments, reported here and elsewhere, bear this out, despite claims of optimality made for the OSP operator.[5,6] The close relationship between OSP and linear unmixing has been discussed in more detail elsewhere.[4,6]

5 SELECTION OF END-MEMBERS OR BASIS VECTORS

Linear unmixing is often applied to mineralogical surveys and plant classification.[7,8] In such applications, the end-members are most often either generated off-line from separate measurements or determined from the image globally. However, such an approach is neither necessary nor desirable for ATD applications, with the exception of the known component of the target spectrum. The background end-members are used for suppression of clutter, and maximum performance is obtained when only those end-members present locally are used within each region of the image. Such an approach also facilitates real-time processing, which is usually required for ATD applications.

Our preferred method for obtaining the background end-members, therefore, is to estimate them separately using a small number of image samples drawn from the near surroundings of each pixel. However, care must be taken to avoid inclusion of the target signature in the background end-member set, when it is present. Experience has shown that for moderately cluttered rural scenes observed at signal-to-noise ratios of a few hundred to one, there are typically less than ten detectable spectral end-members. Therefore, when targets are marginally resolved, use of the second or third nearest neighbors of the pixel of interest provides an ideal data set with which to estimate the background end-members.

Many authors have used a form of principal component analysis (basically, singular value decomposition)[9] to reduce the dimensionality of the data set before attempting to determine the end-members.[10] Principal components are generally linear combinations of true end-members that are mutually orthogonal and sorted in order of their importance for explaining the covariance of the background set. The principal component analysis allows the optimal number of end-members that should be used to model the background to be determined. False alarms from the χ^2 filter are increased whenever too few or too many end-members are used in the fit.

Once the optimal number of end-members has been determined, estimation of principal components can be accomplished without *a priori* information by using the convexity constraint, which relates the observed pixel spectra to those of the end-members.[11] The convexity constraint requires that the coefficients of all of the end-members that make up the background pixel spectra be positive and that they sum to 1. The end-members can be estimated by finding the smallest simplex in the subspace of the principal components that contains all of the background pixel spectra. Unfortunately, known methods for achieving this estimation are computationally intensive and somewhat inexact.

Pending development of an efficient method for extracting end-members from the principal components of the background pixels, we have adopted a less powerful technique to carry out our linear unmixing experiments. The method is simple and intuitively appealing, although it does not take advantage of the information contained in the convexity constraint and is therefore sub-optimal.

We have investigated two methods to derive a set of sub-optimal background basis vectors to be used in an unconstrained linear fit. The first method involves projecting the pixel of interest onto the subspace of the principal components, yielding their coefficients directly.

The second method is simpler and also produces good results. It is based on the following logic: inasmuch as the background pixels are all random linear combinations of the end-members; the pixel of interest, if it does not contain the target, can be modeled as a linear combination of nearby background pixels. This will be true whenever the number of pixels in the background set exceeds the number of end-members present in the image locally. Results can be improved by removing those background pixels with coefficients of low significance and iterating the linear fit until convergence is obtained.

6 AUTOMATED TARGET DETECTION RESULTS

Figure 1 shows an example of a hyperspectral image containing real targets. The scene consists of vehicles parked on a road in the desert. A subset of the original HYDICE image has been degraded in signal-to-noise ratio (SNR) and spatial resolution in order to provide a stressing scenario for algorithm performance evaluation. Forty spectral bands in the near infrared, selected so as to avoid strong atmospheric absorption lines, were used for processing. The signal to noise in the brightest band was degraded to 100:1 by addition of zero-mean white Gaussian noise. The spatial resolution was then degraded by means of a 3×3 pixel aggregation such that the dimensions of the four known targets in the image (marked with circles in Figure 1) were approximately 1.3×2.2 pixels. The targets cannot be distinguished from other similar-sized objects within the image, except

Figure 1 *A desert image containing four marginally unresolved targets*

for the fact that they are on the road (but the ATD algorithm does not, of course, rely on this information).

Results from ATD using the OSP filter and the $\chi^2 K$ filter are shown in Figure 2. A fixed number of background basis vectors consisting of averaged pairs of second-nearest neighbors was used in the filter computations. To assess filter performance, the threshold has been set equal to the smallest target pixel output value, and the total number of spatially distinct false alarms was tabulated. The $\chi^2 K$ filter, shown as blue crosses, produces one false alarm over the $102 \times 106 \times 40$-band image. The OSP filter produces four additional isolated instances of false alarms. The performance of the $\chi^2 K$ filter in this image is solely attributable to K, because use of the K filter alone would have produced the same number of false alarms. This example is typical of results obtained previously using synthetic targets in desert images, except that previous simulations show significant performance from the use of χ^2, especially at lower signal-to-noise ratios. The predominant role of K in these images is because of the relatively high SNRs used for processing (the SNR is three times higher after pixel aggregation) and also because the target signatures, drawn from interior target pixels in the high-resolution images, are artificially well known. Our previous work using synthetic targets had signatures that were only 50% known. The result illustrates the fact that performance of the $\chi^2 K$ filter in desert scenes is generally excellent, even at modest SNRs. Further improvements in performance are possible by using a more sophisticated strategy for selecting the basis vectors used in the linear fits, as discussed above. It is likely that such improvements would enable good performance at significantly lower SNRs at the

Figure 2 *Example of ATD algorithm performance at an SNR of 100:1 in the desert*

expense of greater processing. At an SNR of 400:1 in the brightest band before pixel aggregation, both filters perform perfectly in this scene, producing zero false alarms.

Figure 3 shows a subset of a hyperspectral image from a forested region. The image was degraded in an identical manner to that described for the desert scene above, except that the final SNR was 400:1 before aggregation of pixels. The targets in this image are approximately 1.7×3.2 pixels in size. The image is shown in the brightest spectral band, near 1 μm. The targets are virtually invisible against the background at optical wavelengths at this resolution.

Figure 4 shows the results of our ATD processing. The $\chi^2 K$ filter again outperforms the OSP operator. Only one false alarm is obtained in this 102×130-pixel image. Even though the target signature is artificially well known in our simulation, we view these results as quite promising and, given their consistency with our earlier simulations using synthetic targets, indicative of actual performance obtainable with a fielded system. Indeed, better results can probably be obtained at the expense of more computation, since little was done to optimize the performance of the filter in this initial investigation using real targets. No attempt was made to fit the optimum number of basis vectors to each pixel's spectrum, and information available from convex geometry theory was not used to constrain the fit.

Figure 5 shows the forested region of Figures 3 and 4. The image was degraded in an identical manner to that described above, except that the final SNR was 250:1 before aggregation of pixels. The number of false alarms has increased from one to two in this case.

Figure 3 *A forested scene containing four marginally unresolved vehicles at an SNR of 400:1*

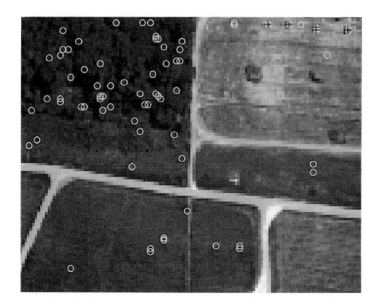

Figure 4 *Example of ATD algorithm performance against forested image containing four targets at an SNR of 400:1*

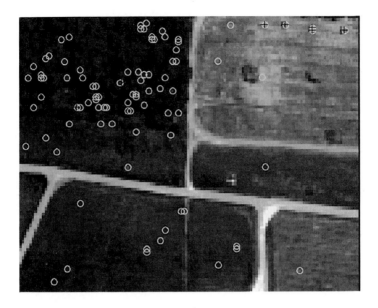

Figure 5 *Example of ATD Processing for a forested image at an SNR of 250:1*

7 CONCLUSION

The results of this study using real targets in high-resolution hyperspectral images bear out the conclusions of our previous work based on synthetic targets embedded in low-resolution hyperspectral images. These conclusions are as follows:

1. The performance of the $\chi^2 K$ filter is superior to that of the OSP filter.
2. ATD of isolated, marginally unresolved, manmade objects in desert scenes is possible using fewer than 40 bands at SNRs near 100:1 or below with false alarm rates on the order of 10^{-4} per pixel or lower.
3. ATD of isolated, marginally unresolved, manmade objects in vegetated scenes is more difficult than for desert scenes but still appears feasible, and SNR requirements are somewhat higher in this case.
4. Performance of the $\chi^2 K$ filter shows promise using a fixed number of nearby background pixels as basis vectors for the linear least squares fitting procedure and would likely be significantly improved using either an adaptive linear fit or true end-members deduced from the local background samples using principles of convex geometry.

References

1. M. R. Descour *et al.*, 1997, *Proceedings of the 1997 Meeting of the IRIS Specialty Group on Passive Sensors*, **1**, 401.
2. R. W. Basedow, D. C. Carmer, and M. E. Anderson, 1995, *SPIE Proceedings*, **2480**, 258.

3. W. F. Kailey and L. Illing, 1996, *SPIE Proceedings*, **2819**, 15.
4. W. F. Kailey and L Illing, 1997, *Proceedings of the 1997 IRIS Specialty Group on Passive Sensors*, **1**, 415.
5. J. C. Harsanyi, W. H. Farrand, and C-I. Chang, 1994, *ACSM ASPRS Annual Convention And Exposition 1994, Reno, NV*, **1**, 236.
6. J. C. Harsanyi and C-I. Chang, 1994, *IEEE Transactions on Geoscience and Remote Sensing*, **32**, 779.
7. M. O. Smith, P. E. Johnson, and J. B. Adams, 1985, *Proceedings of the 15th Lunar and Planetary Science Conference, Part 2, Geophysical Research*, **90**, suppl., C797.
8. A. R. Huete, 1986, *Remote Sensing of the Environment*, **19**, 237.
9. W. H. Press *et al.*, 'Numerical Recipes', Cambridge University Press, New York, 1986.
10. A. A. Green, M. Berman, P. Switzer, and M. D. Craig, 1988, *IEEE Transactions on Geoscience and Remote Sensing*, **26**, 65.
11. J. W. Boardman, 1993, *Summaries of the Fourth Annual JPL Airborne Geoscience Workshop*, **1**, 11.

Jim Janesick, Jeff Pinter, Jim McCarthy and Taner Dosluoglu

Pixel ■ Vision
4952 Warner Avenue, Suite 300
Huntington Beach, CA 92649
mypixel@AOL.com

1 INTRODUCTION

Ideally, three primary mechanisms are responsible for charge transfer in CCDs: thermal diffusion, self-induced drift, and a fringing field effect. The relative importance of each of these is primarily dependent on the charge packet size. Both thermal diffusion and fringing fields are important in transferring small charge packets whereas self-induced drift, caused by mutual electrostatic repulsion of the carriers within a packet, will dominate charge transfer for large packets.

Charge transfer can be summarized as follows for a multi-phase CCD. We begin with a full well of charge under a collecting phase. Next, a neighboring barrier phase is switched instantaneously to be a collecting phase. The main force that causes charge to transfer into this phase is by electrostatic repulsion (i.e., self-induced drift). Charge will divide equally between the two high phases. Shortly there after (i.e., one clock overlap period) the original collecting phase becomes a barrier phase. Again, electrostatic repulsion begins the transfer process. However, as charge density decreases, the self-induced drift field is less effective. The remaining charge is transferred either by fringing fields, which are strongest near the edges of the phase, or by thermal diffusion for that charge that is located near the center of the phase.

Figure 1 presents PISCES modeling data in transferring a charge packet between two Gates. A 5 V potential difference is assumed between the phases. At t = 0 the charge packet is under the left-hand gate. In 10 ps 31% of the charge is transferred to the neighboring phase mainly by self-induced drift. In one nsec 93% of the charge is transferred. At this point charge will be transferred by diffusion because self-induced drift and fringing fields have completed their job.

Nearly all charge transfer efficiency (CTE) speed problems encountered are associated with the rise and fall time of the applied clocks. In fact, if the slew rate of the clocks had infinite speed the CCD would not function! As will be described below, the charge diffusion problem is solely dependent on the falling edge of the clock. The substrate bounce problem, discussed in Section 4, is associated with the rising edge of the clock. Overall, the slower the rise and fall time of the clocks the better the CTE performance. Also keep in mind that all high-speed CTE speed problems discussed below will degrade the sensor's charge capacity and dynamic range. Other CCD performance parameters are for the most part unaffected. These points will become clearer as we go along.

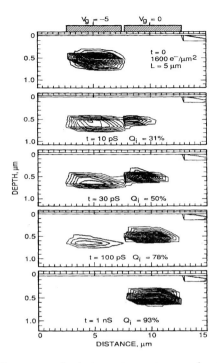

Figure 1 *Charge transfer between two gates as a function of time*

Unless otherwise indicated, data presented in this paper is based on the following three-phase clock wave-shaping formula:

$$\tau_{WS} = \frac{T_T}{12} \qquad (1)$$

where τ_{WS} is the RC clock wave-shaping time constant (s), and T_T is the length of time to transfer one line vertically (T_L) or one pixel horizontally (T_P). As shown in Figure 2, two time constants are given to each clock-overlap period. Twelve time constants are required for a complete line or pixel transfer cycle (T_L or T_P).

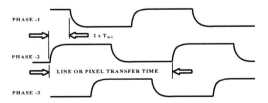

Figure 2 *Three-phase timing diagram showing required clock overlaps and wave shaping*

Example 1

Determine the wave-shaping time constant for a line transfer period of 0.60 μsec.

Solution:

From Equation 1,

$$\tau_{WS} = (6 \times 10^{-7}) / 12$$

$$\tau_{WS} = 5 \times 10^{-8} \text{ s}$$

2 CHARGE TRANSFER MECHANISMS

2.1 Thermal Diffusion

Charge diffusion limits transfer rate when self-induced and fringing field drift are not present. The fractional charge that remains under a phase after transfer time t when diffusion applies is,[1]

$$CTI_D = e^{-t / \tau_{th}} \tag{2}$$

where the time constant, τ_{th}, is,

$$\tau_{th} = \frac{L^2}{2.5 \, D_N} \tag{3}$$

where L is the gate length (cm) and the diffusion constant is given as,

$$D_N = \frac{kT}{q} \, \mu_{SI} \tag{4}$$

where μ_{SI} is electron mobility (cm^2/V-s).

Mobility changes slightly with operating temperature caused by lattice scattering of electrons. The effect is due to the thermal vibration of the atoms of the crystal lattice that disrupts the periodicity of the lattice and impedes the motion of electrons. Decreasing operating temperature reduces lattice scattering increasing mobility. However, the overall diffusion time will increase as temperature is lowered because of kT.

Example 2

Determine the diffusion time constant and the time required to achieve CTE = 0.9999. Assume L = 10 μm.

Solution:

From Equation 4,

$$D_N = 0.0259 \times 1350$$

$$D_N = 35 \text{ cm}^2/\text{s}$$

From Equation 3,

$$\tau_{th} = (10 \times 10^{-4})^2 / (2.5 \times 35)$$

$$\tau_{th} = 1.14 \times 10^{-8} \text{ s } (11.4 \text{ nsec})$$

10 times constants are required to reach a CTE = 0.9999 ($CTI_D = 10^{-4} \cong e^{-10}$) or a total time period of 114 nsec.

2.2 Self-induced Drift

The fractional charge remaining after transfer time t due to self induced drift is,[2]

$$CTI_{SID} = (1 + \frac{t}{\tau_{SID}})^{-1} \tag{5}$$

where τ_{SID} is,

$$\tau_{SID} = \frac{2 \; L^2 \; C_{EFF}}{\pi \; \mu_{SI} \; q \; N_O} \tag{6}$$

where N_O is the number of electrons per unit area (e-/cm^2) for the charge packet at t=0 and C_{EFF} is the effective storage capacitance (F/cm^2).

Note that transfer time has a hyperbolic time dependence because the field strength decreases as charge is transferred (i.e., the process is not exponential as is charge diffusion and fringing fields). The self-induced field decreases until thermal diffusion fields take over. The time required to reach this condition is approximately equal to the diffusion time constant, τ_{th}.

Example 3

Calculate the time required to transfer 90 % of the charge by self-induced drift. Assume L = 10 µm, $C_{EFF} = 2.05 \times 10^{-8}$ F/cm^2 and $N_O = 8 \times 10^{11}$ e- /cm^2.

Solution:

From Equation 6,

$$\tau_{SID} = 2 \times (10^{-3})^2 \times (2.05 \times 10^{-8}) / (3.14 \times 1350 \times (1.6 \times 10^{-19}) \times (8 \times 10^{11}))$$

$$\tau_{SID} = 7.55 \times 10^{-11} \text{ s}$$

Solving for t in Equation 5,

$$t = \tau_{SID} (1/CTI - 1)$$

$$t = (7.55 \times 10^{-11}) \times (1/0.01 - 1)$$

$$t = 7.55 \times 10^{-9} \text{ s}$$

Note that most of the charge is transferred by self-induced drift compared with thermal diffusion. If no fringing fields exist, diffusion becomes dominant over self-induced drift in approximately 10^{-8} s.

2.3 Fringing Field Drift

The fractional charge remaining after transfer time t due to fringing field drift is,[2]

$$CTI_{FF} = e^{t / \tau_{FF}} \tag{7}$$

where τ_{FF} is the fringing field time constant,

$$\tau_{FF} = \frac{L}{2 \, \mu_{SI} \, E_{min}} \tag{8}$$

where E_{min} is the minimum electric field strength under the gate,

$$E_{min} = \frac{2.09 \, \Delta V_G \, \varepsilon_{SI}}{L^2 \, C_{EFF}} \tag{9}$$

where ε_{SI} is the permitivity of silicon (1.04×10^{-12}) and ΔV_G is the potential difference between gates (V).

Example 4

Calculate the fringing field time constant to yield CTE = 0.9999. Assume L = 10 μm, $C_{EFF} = 2.05 \times 10^{-8}$ F/cm² and $\Delta V_g = 5$ V.

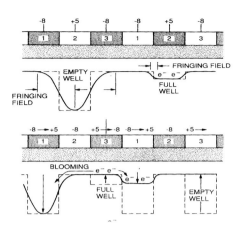

Figure 3 *Fringing fields between collecting and barrier phases for an empty well and full well*

Solution:

From Equation 9 the minimum fringing field generated is,

$$E_{min} = 2.09 \times 5 \times 1.04 \times 10^{-12} / (10^{-6} \times 2.05 \times 10^{-8})$$

$$E_{min} = 530 \text{ V/cm}$$

The corresponding time constant is found from Equation 8,

$$\tau_{FF} = 10^{-3} / (2 \times 1350 \times 530)$$

$$\tau_{FF} = 6.98 \times 10^{-10} \text{ s}$$

10 time constants yields CTE = 0.9999 or a total time of 6.98×10^{-9} s.

This example shows that fringing fields are dominant over diffusion fields for small gates (< 15 μm). However, fringing field strength is highly dependent on signal level and becomes nearly nonexistent at full well. This is because the potential difference between the collecting and barrier phases is considerably smaller at full well. Under full well conditions charge diffusion plays an important role in the transfer process. Figure 3 illustrates the problem. The top drawing shows much stronger fringing fields when a potential well is empty. The bottom figure illustrates the diffusion blooming problem when the falling edge of the clock is too fast. Charge that remains in a collapsing phase must primarily transfer by thermal diffusion.CTI as a function of transfer time at different signal levels is modeled in Figure 4a. Note that transfer time increases as charge density increases. The effect is enhanced when the gate length is made longer. Figure 4b plots CTI for various gate lengths

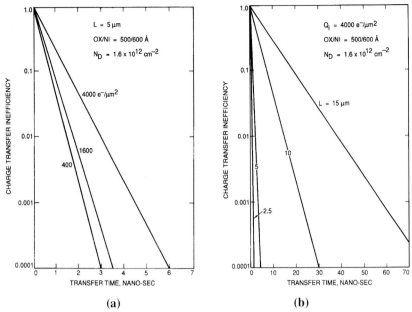

Figure 4 *(a) Charge transfer inefficiency as a function of time and charge density, and (b) charge transfer inefficiency as a function of time and gate length*

assuming a charge density of 4000 e-/μm^2. Note that CTI increases with gate length because the diffusion time constant increases by the square of gate length (Equation 3).

Figure 5a presents full well transfer data generated by a three-phase 1024 × 1024 CCD. The family of curves were taken at different line transfer periods (T_L = 1.0, 0.86 μsec). For line transfer periods < 1 μsec full well degrades in part to the charge diffusion effect on the falling edge of the clock. Wave shaping was employed using Equation 1. For example, at T_L = 1 μsec, τ_{WS} = 8 × 10^{-8} s.

The full well transfer curve in Figure 5b shows that very high transfer rates are possible if the signal level is significantly below full well. For example, Figure 5b shows a Fe^{55} X-ray transfer response at T_L = 200 nsec. Fringing fields are primarily responsible for transferring charge at this signal level (i.e., 1620 e-). PISCES models predict that a T_L < 50 ns is required to transfer charge packets this size. Unfortunately, generating three-phase clocks this fast is difficult to verify experimentally.

Full well data above were taken from the vertical registers. It is important to point out that the horizontal register exhibits the same speed characteristics, in that the gate lengths are the same. However, for most CCD applications the horizontal register must be clocked at a much higher rate. For this reason, the horizontal register is made much wider to accommodate a vertical full well signal while still maintaining fringing fields and high speed. In other words, the horizontal register shifts charge levels well below its own full well level. Horizontal speed (i.e., pixels/s) is usually specified at the vertical full well level.

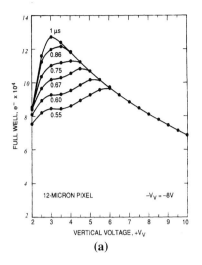

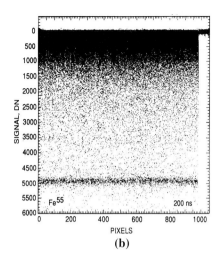

(a)   (b)

Figure 5 *(a) Full well transfer as a function of line time, and (b) X-ray transfer for $T_L = 200$ nsec of line time*

3 CLOCK PROPAGATION

The discussions above show that charge motion was ideally limited by thermal diffusion, self-induced drift, and fringing field drift. Charge transfer can also be limited by a transfer effect called "clock propagation." Clock propagation is associated with the poly resistance and pixel capacitance of the CCD. The impedance will wave-shape the applied clock as it works its way to the center of the array. This effect is therefore dependent on the width of the clock and not the edges. Ideally, the wave-shaping network at the output of the clock driver should set the wave-shaping time constant, $\tau_{WS,}$ and override other wave-shaping components related to the CCD. However, as the clock frequency is increased the clock widths must decrease, and the self-induced wave-shaping effect will become more significant, especially in the center of the array. Eventually the clock amplitude will drop, and, in turn, degrade charge capacity as dictated by full well transfer. The amount of wave shaping that takes place can be estimated by the following formula:

$$\tau_{POLY} = 1.5 \ R_{PIX} \ C_{PIX} \ N_{PIXC}^2 \qquad (10)$$

where τ_{POLY} is the clock time constant at the center of the array (s) and N_{PIXC} is the number of pixels to the center. C_{PIX} is the drive capacitance associated with a pixel, which includes oxide, depletion, and overlap capacitance (F/pixel). R_{PIX} is the resistance of a pixel (ohms/pixel) associated with the poly resistance and is given by the following:

$$R_{PIX} = R_S \ \frac{L_{PIX}}{W_{PIX}} \qquad (11)$$

where R_S is the sheet resistance of the poly layer (ohm/square), L_{PIX} is the length of a phase (cm), and W_{PIX} is the width of the phase (cm). Equation 11 assumes that the vertical registers are driven from both sides of the array.

Example 5

Estimate the poly propagation wave shaping at the center of a three-phase $1024 \times 1024 \times 12$-$\mu$m pixel CCD. Assume $R_S = 40$ ohm/square and a drive capacitance of 10^{-8} F/cm^2. Also determine the minimum line transfer time, T_L.

Solution:

From Equation 11,

$$R_{PIX} = 40 \times (12 \times 10^{-4})/(4 \times 10^{-4})$$

$$R_{PIX} = 120 \text{ ohm/pixel}$$

The capacitance associated with one phase of a pixel is,

$$C_{PIX} = (12 \times 10^{-4}) \times (4 \times 10^{-4}) \times (10^{-8})$$

$$C_{PIX} = 4.8 \times 10^{-15} \text{ F/pixel}$$

From Equation 10,

$$\tau_{POLY} = 1.5 \times 120 \times (4.8 \times 10^{-15}) \times 512^2$$

$$\tau_{POLY} = 0.227 \text{ }\mu\text{sec}$$

From Equation 1,

$$T_L = 12 \times (2.27 \times 10^{-7})$$

$$T_L = 2.7 \text{ }\mu\text{sec (12 time constants)}$$

Frame-transfer CCDs are especially sensitive to poly wave shaping because the vertical registers are clocked as fast as possible to avoid image smearing. The line transfer time required for frame transfer operation is the following:

$$T_L = \frac{S_M \, T_I}{N_L} \tag{12}$$

where T_L is the line time (s), S_M is the smear allowed, T_I is the integration time (s), and N_L is the number of lines transferred into the storage register.

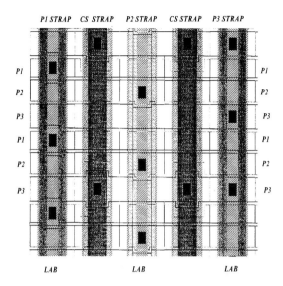

Figure 6 *High-speed backside-illuminated design showing poly and channel stop metalstrapping buss lines*

Example 6

Determine T_L for a frame-transfer $1024(V) \times 512(H)$ CCD. Assume that the integration time is 1/30 s with a desired smear of 1 %.

Solution:

From Equation 12,

$$T_L = 0.01 \times 0.033/512$$

$$T_L = 0.644 \ \mu sec$$

The example above shows that the required line time may not be realizable, because of the clock propagation effect. Strapping the poly gates with metal buss lines that run vertically down the array can significantly reduce poly resistance. For example, Figure 6 shows a high-speed backside-illuminated CCD design that employs poly metal strapping. The metal straps are contacted to the poly phases over the channel stop regions.

It should be mentioned that metal busses that lead up to the poly gates could add to clock propagation. Some CCD manufacturers employ refractory metals that exhibit high impedance. Even aluminum buss lines may limit speed for large CCD arrays, depending on the buss width and drive capacitance. Impedance calculations are performed to specify the buss widths for the maximum clock rate anticipated.

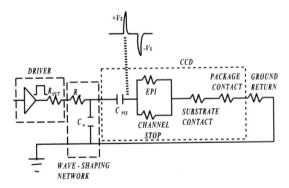

Figure 7 *Substrate return impedances*

4 SUBSTRATE BOUNCE

Substrate bounce is an important mechanism that can affect high-speed operation. In fact, the problem is currently the primary limitation of speed for backside-illuminated CCDs. Bounce is associated with the sensor's ground return. As the CCD is clocked, displacement currents must flow in and out of the device through substrate return. For an n-channel CCD, this current takes the form of holes that originate from the p-depletion and gate inversion regions as they collapse and form during charge transfer. If a ground impedance exists, an IR drop is generated, causing the substrate to "bounce" in potential. The rising edge of a clock causes the substrate to bounce positively, because holes are trying to leave the CCD. The falling edge produces a negative bounce as holes return to the sensor. In that substrate is the master reference point for all CCD clock and bias potentials; substrate bounce can cause havoc on operating potentials and performance. For example, bounce dramatically affects charge capacity as, demonstrated below. Bounce problems are more prevalent for the vertical clocks because of high displacements currents generated by these registers. Discussions on substrate bounce below refer to a three-phase CCD. However, the problem is common to all CCD families.

Figure 7 shows important impedances related to the ground return. The connection made between the CCD and ground is one problem area that generates bounce. For example, the inductance associated with the ground conductor must be considered at high clocking rates. The epoxy layer between the CCD and metal package can also exhibit bounce problems. Aluminum-backed CCDs mounted with certain conductive epoxies can form a high-resistance insulating layer, resulting in severe bounce problems. Therefore, gold is usually the metal of choice to make a low impedance epoxy ground connection to the CCD and package. Backside-illuminated CCDs require special substrate contacts, because the substrate layer is eliminated through the thinning process, preventing a direct epoxy contact, as with front-side illuminated CCDs. Frequently, a single-point substrate contact is made on the front of the CCD; however, this type of contact often exhibits speed problems because of the finite contact resistance involved. To avoid this problem, the contact area

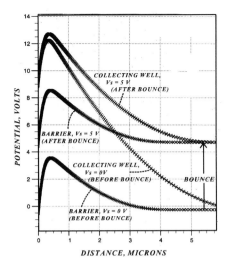

Figure 8 *Potential well plots illustrating the substrate bounce problem*

can be increased or may encircle the entire CCD, providing a lower impedance return. A backside contact can also be made to the backside aluminum light shield for frame transfer layouts. For this purpose, the aluminum must make intimate contact with the silicon for a low impedance return.

The rising edge of the clock generates the most bounce problems, for two reasons. First, charge capacity is less sensitive to negative bounce because two collecting phases exist at that time, compared with a single collecting phase when positive bounce occurs (refer to the timing diagram shown in Figure 2). Figure 8, which shows potential well plots before and after a 5 V bounce occurs, illustrates the second problem. The substrate potential and the surface potential of the barrier phase before bounce are at zero volts (i.e., the barrier phase is inverted and pinned). Note that the surface potential for the barrier phase follows the substrate when a bounce occurs (i.e., both move 5 V upward). The potential maximum for the collecting phase only shifts up slightly. Therefore, the potential difference between the collecting and barrier phases ($V_C - V_B$) decreases by approximately 5 V, which causes a large full well reduction, as dictated by full well transfer.

Referring back to Figure 2, note that the rising edge of Phase-1 will cause Phase-2 to bounce, causing Phase-3 to bloom. Likewise, the rising edge of Phase-3 will cause Phase-1 to bounce, causing Phase-2 to bloom. Finally, the rising edge of Phase-2 will cause Phase-3 to bounce, causing Phase-1 to bloom. Blooming will be more significant for that phase with the lowest amount of charge capacity when bounce occurs.

The amount of substrate bounce generated is (1) proportional to the capacitance associated with a clock phase being stimulated, (2) proportional to the substrate resistance, and (3) inversely proportional to the rise time of the applied clock. If the substrate resistance is external to the CCD, the substrate bounce as a function of time is as follows:

$$V_{SB}(t) = \frac{A_C}{1 - \tau_{WS}/\tau_{SB}} \ (e^{-t \ / \ \tau_{SB}} - e^{-t \ / \ \tau_{WS}}) \tag{13}$$

where A_C is clock amplitude with a wave-shaping time constant of τ_{WS} of,

$$\tau_{WS} = R_W \ C_W \tag{14}$$

where R_W and C_W are the resistance and capacitance values of the clock wave-shaping network shown in Figure 9 and τ_{SB} is the substrate bounce time constant,

$$\tau_{SB} = R_{SB} \ C_B \tag{15}$$

where R_{SB} is the external substrate impedance (ohms), and C_B is the capacitance related to the barrier phase that is inverted (cm^2). That is,

$$C_B = A_B \ C_{OX} \tag{16}$$

where A_B is the total array area for the barrier phase (cm^2) and C_{OX} is the oxide gate capacitance (F/cm^2). Note that the capacitance associated with the rising edge of the clock is simply the oxide capacitance, because the barrier phase remains inverted and pinned at all times during the bounce effect. However, it is possible that the barrier phase may come out of the pinned condition on the falling edge. In this case the drive capacitance will be considerably less, because the potential well depletion capacitance will be in series with C_{OX}.

Example 7

Using Equation 13, plot substrate bounce as a function of time for clock wave-shaping time constants of $\tau_2 = 10^{-7}$, 10^{-6} and 10^{-5} s. Assume a three-phase $1024 \times 1024 \times 12$-μm pixel CCD, a gate oxide capacitance of 3.45×10^{-8} F/cm^2, and a channel stop width of 4 μm. Also, assume an external substrate resistance of 100 ohms. Neglect poly impedance.

Solution:

From Equation 16,

$$C_B = (3.45 \times 10^{-8}) \times (4 \times 10^{-4}) \times (8 \times 10^{-4}) \times 1024^2$$

$$C_B = 1.2 \times 10^{-8} \ F$$

From Equation 15,

$$\tau_{SB} = 10^2 \times (1.2 \times 10^{-8})$$

$$\tau_{SB} = 1.2 \times 10^{-6} \ s$$

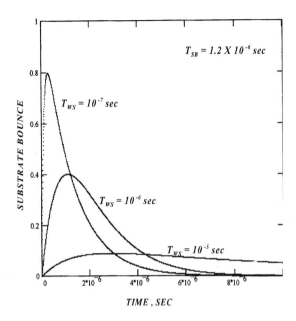

Figure 9 *Substrate bounce for different wave-shaping conditions*

Figure 9 plots substrate bounce for wave-shaping time constants of $\tau_{WS} = 10^{-7}$, 10^{-6}, and 10^{-5} s. Note that the clock wave-shaping time constant must be greater than the CCD time constant to suppress substrate bounce to a negligible level.

Figure 10 is an experimental full well transfer generated by a 1024 × 1024 × 12-μm pixel CCD clocked into inversion. A resistor was inserted between the CCD and ground to purposely cause a substrate bounce problem. Note that as the substrate resistance increases, optimum full well shifts to a higher clock voltage. For $R_{SB} = 1$ k ohm, a shift of 3 V is experienced.

Holes for a backside-illuminated CCD must laterally travel through neutral epitaxial material below the depletion region or channel stops to reach ground return. In either case, the impedance experienced by the holes can be high, generating bounce when clock edges are too fast. It should be noted that CCDs with vertical antiblooming (VAB) exhibit similar bounce problems to those of backside CCDs. VAB is built on n-epitaxial material. The p-well associated with these sensors is completely depleted, except for material beneath the channel stop regions. Therefore, holes must also leave through the channel stops, resulting in substrate bounce.

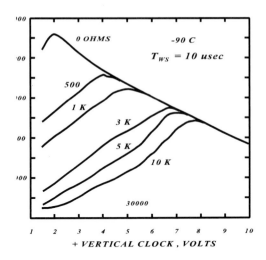

Figure 10 *Full well transfer curves generated with different substrate resistors*

Example 8

Calculate the substrate resistance associated with a single pixel shown in Figure 11.

Region 1 = Lateral channel stop resistance (per pixel)

Region 2 = Resistance from channel stop to the depletion edge (per pixel)

Region 3 = Lateral resistance of backside field-free material (per pixel)

Assume channel-stop resistivity = 1500 ohm/square and epitaxial resistivity = 40 ohm-cm.

Solution:

Region 1:

$$R_{CS} = 1500 \times (12 \times 10^{-4}) / (4 \times 10^{-4})$$

$$R_{CS} = 4.5 \times 10^{3} \text{ ohms/pixel}$$

Region 2:

$$R_{EPI} = 40 \times (6 \times 10^{-4}) / ((4 \times 10^{-4})(12 \times 10^{-4}))$$

$$R_{EPI} = 1.38 \times 10^{4} \text{ ohms/pixel}$$

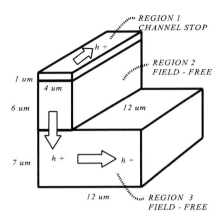

Figure 11 *Cross-section of a pixel showing hole return paths*

Region 3:

$$R_{FF} = 40 \times (12 \times 10^{-4}) / ((7 \times 10^{-4})(12 \times 10^{-4}))$$

$$R_{FF} = 5.7 \times 10^4 \text{ ohms/pixel}$$

This example shows that most hole current for a backside-illuminated CCD flows through the channel-stops.

Substrate bounce increases dramatically when the barrier phases are clocked into inversion, because extra holes must move in and out of the CCD. The full well transfer curves presented in Figure 12 show the effect. Data is taken from a 455(H) × 1024(V) × 12-μm pixel backside-illuminated split-frame transfer CCD using 200 nsec clock wave shaping. Inversion for the CCD occurs at -10 V. Note that full well performance degrades on either side of inversion. When the barriers are set below inversion, charge capacity is reduced because the potential between the collecting and barrier phases is reduced (i.e., $(V_C - V_B)$). Clocking the CCD deeper into inversion increases the substrate bounce problem, again degrading full well. Optimum full well occurs at the onset of inversion without significant bounce. Wave shaping is optimally adjusted for speed and charge capacity.

Figure 13 shows a 7-s dark current column trace for the CCD characterized in Figure 12. The barrier phases are set into deep inversion ($V_B = -15$ V) and wave-shaped at 200 nsec, making the sensor vulnerable to substrate bounce. The first 256 lines read out are from the frame storage section. CTE is well behaved in this region, because the light shield on the backside of the CCD acts as a ground plane, suppressing bounce problems. Holes in this region need only travel down from the channel stops through field-free material to the light shield, a distance of approximately 14 microns for this CCD. This represents the same impedance as a frontside-illuminated CCD. However, within the imaging region (lines

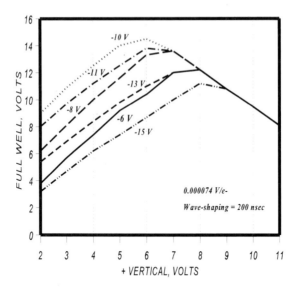

Figure 12 *High-speed full well transfer curves for a backside-illuminated CCD*

257–768), the holes have no direct ground path. In this region, holes must laterally flow down the channel stops towards the storage region. The center pixel of the imaging region sees the highest impedance and experiences the greatest amount of "channel stop bounce." The column trace shown reflects this problem. Approximately 150 lines up from the image/storage interface, the bounce causes the CCD to bloom, destroying dark current pixel

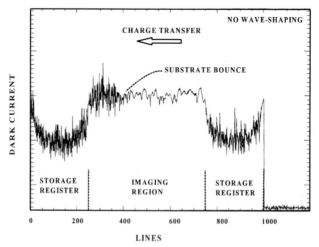

Figure 13 *Dark current column trace showing the channel stop substrate bounce effect without wave-shaping*

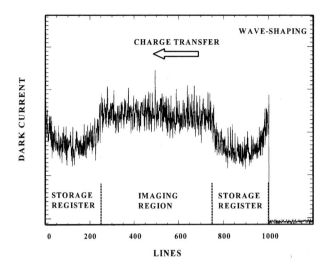

Figure 14 *Improved full well response with wave shaping*

non-uniformity. The second storage region (lines 769–1024) also has a light shield and ground return. However, charge in this region must transfer through the imaging region; therefore, its full well is limited to the same level as the imaging region. Note that the dark current level within the storage region is lower than the imaging region because of accumulation differences caused by the light shield. Figure 14 shows the same response with additional wave shaping. Optimum full well characteristics return, with sacrifice in speed (i.e., the -10 V curve of Figure 13).

Two-phase CCDs are less sensitive to substrate bounce problems than multi-phase operations, because clocking is symmetric. That is, the phases can be clocked 180 degrees out of phase with each another. Under these clocking conditions, holes will be transferred from phase to phase instead of leaving the substrate (holes travel in the opposite direction to electron transfer). Four-phase CCDs can also be clocked in a two-phase fashion, thus reducing substrate bounce. However, there is a full well penalty when clocking a four-phase in this manner. Virtual-phase CCDs are sensitive to substrate bounce in that this technology uses a single clock that must be inverted.

References

1   J. Carnes, W. Kosonocky, and E. Ramberg, *IEEE Trans. Electron Devices*, June 1972, **ED-19**, 798.

2   D. Barbe, *Proc. IEEE*, January 1975, **63**, 38.

Further Reading on High Speed Scientific CCDs

J. Janesick, A. Dingizian, G. Williams, and M. Blouke, in 'Recent Developments in Scientific Optical Imaging', Eds. M. B. Denton, R. E. Fields, and Q. S. Hanley, Royal Society of Chemistry: Cambridge, UK, 1996; pp 1–14.

HIGH-SPEED ARRAY DETECTORS WITH SUB-ELECTRON READ NOISE

Craig D. Mackay

Institute of Astronomy, University of Cambridge
Madingley Road, Cambridge, CB3 0HA, England
tel: +44-1223-337543   fax: +44-1223-330804
e-mail: cdm@ast.cam.ac.uk

1 ADAPTIVE OPTICS: INTRODUCTION

One of the biggest challenges facing astronomers today is the problem of reducing the harmful effects of the atmosphere on the quality of astronomical images formed at telescopes. Today, telescopes have been built with the diameter of 10 m, and a new generation of even larger telescopes is planned. Simple optical diffraction theory shows that a 10-m telescope should have a resolution, in principle, of about 0.01 arcsec. In practice, these telescopes produce images that are only a fraction of 1 arcsec, many tens of times poorer in angular diameter than one would expect from diffraction theory. The reason for this poor performance is that the light propagating through the atmosphere into the telescope is wavefront distorted by phase fluctuations in the atmosphere. These phase fluctuations mean that the image formed is smeared out and substantially reduced in detail. In the visible part of the spectrum, the scale size of fluctuations in the atmosphere is of the order of 10-30 cm, and the time scale over which features in the atmosphere stay constant is of the order of 10-30 milliseconds. These phase variations can be seen easily by taking high-speed images of the light coming into the aperture of telescope. The distortions seen in the image simply reflect the fact that the incoming wavefront has been distorted and rays have been bent away from the original path by a small amount producing interference patterns on the picture. Astronomers would like to be able to measure phase errors at each part of the telescope aperture quickly enough to derive the corrections in time to distort a flexible mirror so as to correct the phase errors accurately.

2 ADAPTIVE OPTICS APPROACHES

The first problem that astronomers encounter is that the short time scales of the fluctuations and the small apertures over which the phase is relatively constant mean that a relatively small amount of light is available to measure phase error. This in turn means that it is only possible to work with bright stars that produce enough light for it to be possible to detect the phase errors rapidly enough. The phase errors in the atmosphere are different looking in different directions in the sky. In practice, this is a very small part of the sky that can be described as having a single-phase measurement. This isoplanatic patch is typically one arc min in diameter. There are a relatively small number of interesting objects that are close enough to bright stars to make natural guide star adaptive optic systems practical. Recently, work has progressed well on the use of laser guide stars to allow better sky coverage. Here, a laser of high power is tuned to the wavelength of sodium atoms. The laser beam is shone up parallel to the beam of the telescope to produce an artificial star by stimulating light emission from the sodium atoms in the upper atmosphere. Even with laser guide stars, the problem of arranging sensors to measure phase fluctuations is quite considerable. Laser guide stars cannot be very bright, because the sodium layer quickly becomes saturated.

Wavefront sensors used at telescopes generally use the Shack-Hartmann sensor arrangement, as shown in Figure 1. This arrangement involves dividing up the apertures of the telescope into small cells and measuring the average phase error across a cell. The pattern of phase errors between cells is then interpolated across the aperture to give the two-dimensional phase correction pattern needed. There is in an amount of light that can land in one of the cells and give a phase measurement accurate enough for phase corrections to be computed. The larger the cell size, the greater will be the signal from the star, but the risk of significant phase errors across a cell, making the measurement inaccurate, will be commensurately great.

In any such adaptive system, the limiting sensitivity is set entirely by the detector used to sense the errors in the wavefront. Traditional charge-coupled devices (CCDs) have good quantum efficiency, but in order to read out the devices at high speed, which is essential for adaptive optic systems, the readout noise inevitably produced markedly limits the sensitivity of the system. Generally, it is necessary to get to frame rates of several hundred frames per second in order to do this successfully. Within each aperture of the Shack-Hartmann lenslet array, we can use simple quadrant autoguiding, leading to characterisation of each cell by a single-phase error.

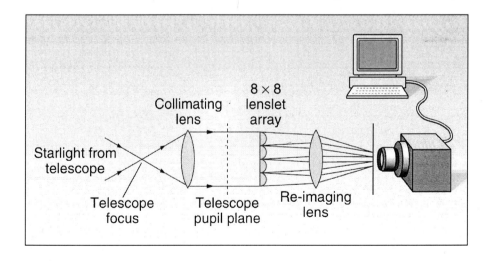

Figure 1 *The layout of the Shack-Hartmann sensor as used as an adaptive optics sensor (Picture courtesy of Laser Focus World)*

3 INFRA-RED ADAPTIVE OPTICS

One way to make the problem more manageable is to not try to do it in the visible, where light is more difficult to come by, and the angular scales of the fluctuations are small, and the temporal scales of the fluctuations are short. We are interested in using a cell large enough so that the phase errors across it are small. Because the phase errors are simply due to refractive index variations in the atmosphere, by moving to the infrared, the phase errors across a given angular scale will become much smaller. In practice, this function goes as $\lambda^{1.6}$. The area of the cell, therefore, goes as $\lambda^{3.2}$. In addition, the correlation times of these atmospheric fluctuations also scales as $\lambda^{1.6}$. Most array detectors, both in the visible and the near infrared, produce one electron for every detected photon. In the infrared, the energy that is emitted corresponds to more photons than it does in the visible so there are more infrared photons for a given energy, in proportion to λ^{1}. The consequence of this is that the total number of photons detectable in an atmospheric cell in one correlation time increases, as one moves towards the red, by a factor of $\lambda^{4.6}$. This clearly makes it very attractive indeed to do this work in the infrared rather than the visible.

4 LOW-NOISE INFRARED ADAPTIVE OPTICS DETECTOR COMPONENTS

Until recently, detectors available in the near infrared were noisy and had relatively low quantum efficiencies. Recent developments, principally by Rockwell in California, have produced a series of near-infrared devices with high quantum efficiency (up to 80% at 2.5 microns) and with a wide sensitivity range from 0.8–2.5 microns wavelength. Devices such as the HAWAII device, with 1024 x 1024 pixels, are able to operate with good readout noise (as low as 8 elections RMS) and may be read at relatively fast pixel rates (several hundred kHz). These devices have now been used widely by astronomers for direct imaging, and they can certainly be used to take fast images on a small part of the device area for adaptive optics applications. The structure of these devices is rather different from the much better known CCDs used for general scientific imaging at optical telescopes. These near-infrared devices are X-Y addressable CMOS devices that allow fast readout of sub arrays with none of the transfer overheads or shuttering problems associated with charge-transfer devices.

5 YET LOWER READ-OUT NOISE

Even for the new HAWAII chips, readout noise is still a serious problem. Further reduction would be helpful in order to improve limiting sensitivity of the detector system. Rockwell have also recently developed a new kind of near-infrared imaging detector that has 128 x 128 pixels each 40 microns square. In the HAWAII devices, the individual cells contain three transistors. One of these transistors connects the integrating capacitor in the common video line and the other two respond to electronic signals from shift registers at the edge of the device that allow an individual pixel to be selected and read out non-destructively. The new high-speed 128 x 128 devices have much bigger pixels, and so it is possible to incorporate within each pixel a high-gain operational amplifier. Because of the presence of this amplifier, it is possible to construct a pixel so that its read noise is on the order of 0.7 elections, even at the very high readout rate of 10 MHz. One recently fabricated device has 32 parallel outputs and therefore is capable of being read out at about 10,000 frames per second with full resolution while still giving excellent noise performance. For most astronomical applications, a frame rate of the order of 1000 frames per second would be perfectly satisfactory, and at those rates, the read noise in these devices will be significantly reduced, possibly to about 0.3 elections RMS.

With that level of readout noise, there is no disadvantage in reading out the chip quickly, in contrast with the situation with CCD-based systems. Individual photons may be detected with good signal-to-noise, so the detector has effectively the same system efficiency as a true photon-counting detector, but in this case with the sort of quantum efficiency obtained from solid-state devices rather than from a photocathode-based device. The ideal means of operating an adaptive optic system in the future using this type of device may well be different from that used to date. Rather than using a Shack-Hartmann sensor, in which there are fixed cell sizes, effectively defined in the hardware as the scale over which atmospheric phase fluctuations can be measured, there are big advantages in using a different kind of wavefront sensor, a shearing interferometer. This would allow a full two-dimensional image to be taken.

The excellent noise performance of these detectors allows for both spatial and temporal over-sampling. The system can then average over spatial and temporal scales to match the exact (and often rather variable) conditions that pertain while the observation is being made. In this way, it is believed that significant improvements will be achieved in the limiting sensitivity obtainable with adaptive optical systems on the latest generation of telescopes.

IMAGING AND SPECTROSCOPY WITH PROGRAMMABLE ARRAY MICROSCOPES

Q. S. Hanley, P. J. Verveer, and T. M. Jovin

Department of Molecular Biology
Max Planck Institute for Biophysical Chemistry
Am Faßberg 11
D-37077 Göttingen, Germany

1 INTRODUCTION

The distinctive feature of a programmable array microscope (PAM) is the placement of a programmable spatial light modulator (SLM) in an image plane of a microscope to create patterned illumination and/or detection. In this arrangement, the SLM may be used to create optical sectioning conditions, to form a spatial encoding mask, or to define regions of interest. This paper covers two types of programmable array microscope. In the first type, patterns are used for both illumination and detection, creating an optical sectioning condition. The degree of optical sectioning depends on the periodic structure of the pattern, the numerical aperture of the objective, the wavelength of light, and the magnification. The second type of microscope uses the spatial light modulator to perform Hadamard spatial encoding in the image plane, which allows the encoded information to be recovered and measured with a CCD camera after the light has passed through an imaging spectrograph. The end result is a stack of images, each of which corresponds to a particular wavelength.

1.1 Optical Sectioning Overview

Commonly available optical sectioning microscopes generate images by using single point scanning, line scanning, or disk scanning.[1-4] Disk scanning systems typically use repeating patterns of dot or line arrays on a disk. More recently, however, systems have been described in which either the conventional Nipkow-type disk is replaced with a disk on which random bit sequences have been etched or the regularly spaced apertures are covered with microlenses.[5,6] The advantage of these methods is that they increase the throughput of the rotating disk and hence the speed of measurement. In this context the programmable array microscope is advantageous because it allows most of these methods to be reproduced under programmable control.

The first programmable array microscope for fluorescence was implemented using a digital micromirror device (DMD).[7,8] This system was used in an arrangement in which the pixels of a CCD camera and the pixels of the SLM were used to establish the

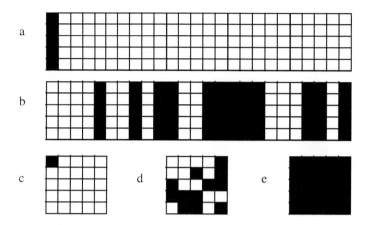

Figure 1 *Patterns of illumination and detection in a PAM. (a) Line scan pattern with a 1 × 25 unit cell. (b) One-dimensional random bit pattern based on a S-matrix sequence of length 31 tiled into a 1 × 25 unit cell. (c) Dot lattice pattern with a 5 × 5 unit cell. (d) Two-dimensional random bit pattern based on an S-matrix sequence of length 31 tiled into a 5 × 5 unit cell. (e) Conventional homogenous illumination. Blacked positions represent "on" pixels*

illumination and detection apertures (Figure 2). This arrangement had two paths, one for illumination and another for detection. The sectioning ability of the system when observing biological specimens was good (Figure 3), but required extensive post-processing. Dual path optical sectioning configurations are more difficult to align than single path arrangements. In a PAM, the latter can be implemented by using the same

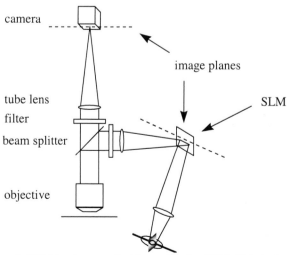

Figure 2 *Dual path PAM implemented with a reflective SLM. This is the geometry of the first fluorescence PAM*

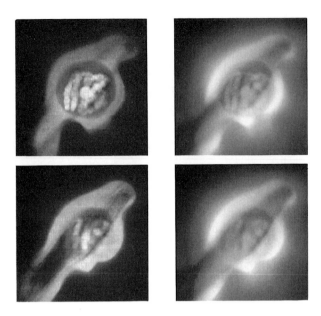

Figure 3 *Optically sectioned images taken with the first fluorescence programmable array microscope. Images on the left show two focal planes in a specimen of Chironimus thumii when the microscope was used in a sectioning mode. Images on the right are the corresponding conventional images*

SLM for both illumination and detection. The limitations of the first generation optical sectioning PAM led to the construction of the single path system, which is one of the topics of this chapter.

1.2 Two-dimensional Imaging Spectroscopy

It is highly desirable to measure spectra in a microscope in an efficient manner. There are three standard ways in which two-dimensional spectral images are acquired: slit scanning, sample scanning, or wavelength tuning (Figure 4). These three methods comprise the bulk of imaging spectroscopy done in microscopes. In situations in which only a limited area of the sample is of interest or a limited number of wavelengths are needed, they remain the methods of choice. However, they share a common disadvantage, in that the majority of the light is rejected or the majority of the sample is outside the region of view. When large areas of a sample are of interest, multiplexed methods allow a greater amount of the sample to be observed simultaneously (Figure 5). Fourier encoding of the wavelength domain has proved successful for generating two-dimensional spectral images. Hadamard transform techniques, however, have a number of special advantages for this type of measurement. When implemented with a stationary mask technology, such as a liquid crystal display (LCD) device, no moving parts are required. The optics are relatively simple and do not require exceptional care to align.

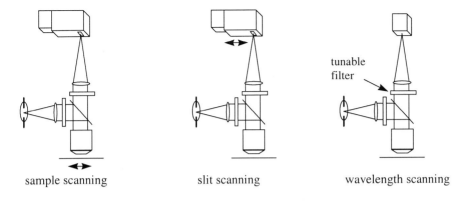

sample scanning   slit scanning   wavelength scanning

Figure 4 *Commonly used methods for generating two-dimensional spectral images.*

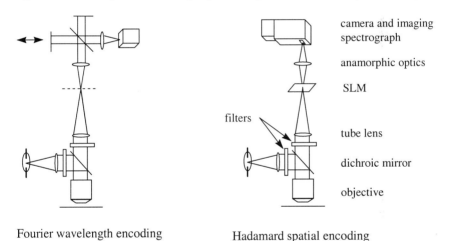

Fourier wavelength encoding   Hadamard spatial encoding

Figure 5 *Two multiplexed methods for generating two-dimensional spectral images.*

The systems (Figures 4 and 5) are found in the commercial and academic literature under many names: imaging spectroscopy, spectral imaging, multispectral imaging, hyper-spectral imaging, digital imaging spectroscopy, and many other variants. We prefer to call methods in which an image with two spatial axes is recorded with a spectrum at every point *two-dimensional spectral imaging*, regardless of the method used to generate it.

Wavelength scanning is currently the most popular of the methods shown in Figure 4 and has been implemented with variable interference filters,[9,10] acousto-optical tuneable filters,[11] and liquid crystal tunable filters.[12] Commercial wavelength scanning systems are available.

Of the encoding methods, Hadamard systems have been more widely reported in the literature.[13-19] However, commercial systems are not available. A Fourier transform system[20,21] applied to fluorescent biological specimens has renewed interest in the field of two-dimensional imaging spectroscopy. A commercial version of this instrument has been applied, most notably, to spectral karyotyping analysis.[22-24]

The attraction of these spectroscopic methods derives from the introduction of multiple absorption and fluorescence probes with specificity for cellular constituents and local physical properties. The limitation to more extensive use of such spectroscopic methods is the high cost and limited availability of suitable instrumentation. A Hadamard transform system incorporating a mass-produced LCD could be used as a simple and low cost alternative to the Fourier methods.

2 EXPERIMENTAL

Two experimental systems have been developed in the course of this research. One is a Hadamard transform spectroscopic system that uses an LCD as the SLM.[25] The second is an optical sectioning system incorporating a digital micromirror device (DMD).[26]

2.1 Optical Sectioning PAM

2.1.1 Instrumentation. The DMD optical sectioning PAM was an add-on to a Nikon E-600 microscope. A Texas Instruments (Plano, TX, USA) DMD was mounted in the primary image plane and used to define patterns of illumination and detection. The light from the "on" elements was relayed to an Apogee KX-2 charge-coupled device (CCD) camera (Apogee Instruments, Tucson, AZ) using 1:1 imaging optics. The quality of the optical relay was sufficient to observe effects from the 1-μm dark segments between adjacent DMD elements. The DMD was illuminated through a Nikon filter block using either a 250 W super high pressure Hg arc lamp system (Lumatec GmbH, Munich, Germany) or a 450 W Xe arc lamp (Müller Elektronik-Optik, Moosinning, Germany). Axial positioning of the objective was accomplished using a PIFOC piezoelectric system with a resolution of 10 nm (Physik Instrumente, Waldbronn, Germany). A block diagram of the microscope and add-on is given in Figure 6.

2.1.2 Illumination and Detection Patterns. Three patterns were studied: conventional illumination, regular 4×6 dot lattices, and 1×24 line patterns. Conventional illumination, consisting of a uniform homogeneous field, was created by setting all elements of the DMD to the "on" position. To generate an image, an image was taken with the specified pattern followed by a blank consisting of an exposure with all elements of the DMD placed in the "off" position.

2.1.3 Axial Resolution Evaluation. The axial resolution of the microscope was measured using ~20-nm-thick rhodamine-B films prepared by spin coating. All data were collected using a Nikon Plan Apo 60× NA 1.4 oil immersion objective. The axial scans of the fluorescent thin films were done with a Hg arc lamp, and a Nikon G-1A filter set consisting of a 546-nm excitation filter with a 10-nm bandpass, a 575-nm dichroic mirror, and a 580-nm-long pass emission filter. These axial scans were sampled every 50 nm.

2.1.4 Biological Specimens. A comparison of the imaging quality obtained using the three pattern types was made using human breast adenocarcinoma cells (MCF7, ATCC HTB22; American Type Culture Collection). The cells were incubated with a monoclonal antibody against tubulin followed by a secondary goat anti-mouse IgG labeled with Oregon green (Molecular Probes, Eugene, OR). Fluorescence was excited

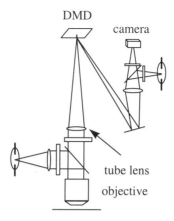

Figure 6 *Block diagram of the optical sectioning PAM*

with the Xe arc lamp and a Nikon BA-2A filter block (450–490-nm excitation filter, 505-nm dichroic mirror, and 520-nm longpass emission filter).

2.2 Hadamard Transform PAM

2.2.1 Instrumentation. A schematic diagram of the fluorescence Hadamard transform programmable array microscope (PAM) is shown in Figure 6. The spectroscopic unit consisted of an add-on to a Nikon E-600 microscope equipped with a 100W Hg arc lamp and an epi-fluorescence unit. This spectroscopy module consisted of a SVGA format LCD device SLM (Central Research Labs, England), anamorphic relay optics, a PARISS prism-based imaging spectrograph (Lightform Inc., Belle Mead, NJ), and a KX-2 CCD camera (Apogee Instruments, Tucson, AZ). The pixels of the LCD consisted of an irregular element approximating a 26-µm (horizontal) by 24-µm (vertical) rectangle, and the active element was a TFT twisted nematic liquid crystal array in an 800 × 600 format. The SLM was used to display a series of bar patterns defined by the rows of cyclic S-matrices. An image was recorded for every row of the S-matrix. After subsequent decoding with an inverse S-transform, a three-dimensional image stack was obtained.

2.2.2 Optical System. The optical system was modeled after a previously described system[14] and consisted of a 260-mm achromatic doublet (Melles Griot) collection lens and two cylinder lenses with 100-mm and 40-mm focal lengths (Melles Griot) arranged perpendicularly. This anamorphic imaging system was designed to relay the light from a 600 × 511 field of SLM pixels through an 8 × 3-mm aperture. The anamorphic optics relayed an ~7.6 × 2.6-mm image to the entrance aperture of the spectrograph. The long axis of the compressed image corresponded to the normal slit axis of the spectrograph and the short axis to the dispersion axis of the spectrograph and shift direction of the Hadamard mask. A ray trace of the optics is shown in Figure 7. After application of the S-transform transformation, the sampling of the primary image plane was ~18 × 33 µm or, upon binning the camera 2 × n, ~37 × 33 µm. It should be noted that binning of the CCD along the y spatial axis has no effect on the image but reduces the sampling density of the spectra. The entrance slit of the spectrograph was removed and the Hadamard

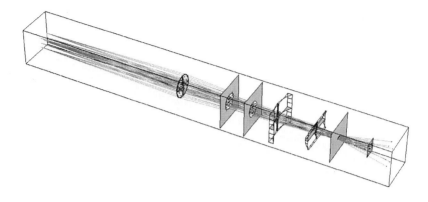

Figure 7 *Ray trace showing arrangement of lenses in the anamorphic relay optics of the Hadamard transform spectroscopic imaging system*

mask patterns used to define a series of entrance slits for the spectrograph. As a result, light entering the spectrograph from each slit was systematically offset from neighboring slits, leading to a systematic shift in the position of the wavelength axis that was linear with position along the transform axis. A wavelength shift correction had to be applied to the data after transformation.

2.2.3 Control of Photobleaching. Specimens can be classified into three categories: 1) those exhibiting negligible photobleaching; 2) those exhibiting moderate photobleaching approximately linear over the set of exposures, and 3) those in which photobleaching is severe. Samples in Category 1 were analyzed using a single set of exposures. Taking two sets of exposures, in the second of which the sequence of masks was reversed, could alleviate the photobleaching of samples in Category 2. The two data sets were then summed and transformed. Under conditions where photobleaching was approximately linear over the set of exposures, this was sufficient to remove bleaching-related artifacts. Samples in Category 3, which are characterized by exponential decay of the signal, may be approached in three ways. The best approach is to prepare the sample using more stable fluorescent probes. A less desirable approach is to pre-expose the sample to the excitation light and intentionally bleach rapidly decaying components so that they to not degrade the spectra from other components. In samples for which this is not an option, the exposures should be broken up into multiple sets. In each set, the exposure time must be minimized such that the photobleaching is approximately linear. The results are then summed to get an image of sufficient signal strength.

2.2.4 Data Acquisition and Processing. Data collection was automated under computer control. At the end of the sequences, a blank was collected consisting of an illuminated sample viewed through the LCD with all the pixels set to the non-transmitting state. This blank was subtracted from each image prior to application of the inverse transform. The data were analyzed using a fast Walsh Hadamard transform algorithm originally written by Norm Mortensen at the University of Kansas. Processing of the data was done with a stand-alone program (MKSPECTRUM), which loads data files from disk, performs background subtraction, sums paired data sets, computes the S-transform, corrects the wavelength axis skewing using either sync or linear interpolation, and writes the processed data to disk.

2.2.5 Biological Specimens A series of two-dimensional spectral images of human breast adenocarcinoma cells (MCF7 ATCC HTB22, American Type Culture Collection) were recorded at 1.0-μm intervals along the z-axis. The cells were stained with MitoTracker™ CMTM Ros-H$_2$ (M7511, Molecular Probes, Eugene, Oregon) and Oregon green. The Oregon green was applied as a goat anti-mouse IgG (Molecular Probes) after the cells had been pre-treated with an anti-tubulin monoclonal antibody resulting in visualization of tubulin. The MitoTracker stained the mitochondria of the cells. Fluorescence was excited at two bands simultaneously with a triple-bandpass filter set (Chroma Technology Corp., Brattleboro, Vermont). This filter set is designed to simultaneously excite at 390–410 nm, 490 nm, and 560–580 nm and detect at 450–470 nm, 510–545 nm, and 600–650 nm. The 390–410 nm channel was blocked using a KV408 longpass filter (Schott Glas, Mainz, Germany). The S-matrix used for encoding was 127×127.

3 RESULTS

3.1 Optical Sectioning PAM

Measured axial responses to rhodamine thin films indicated a resolution of 1.0 μm for line patterns having a 1×24 unit cell and 0.6 μm for the dot lattices having a unit cell of 4×6 (Figure 8). A conventional illumination pattern showed no ability to resolve the thin fluorescent film. The line shape of the axial response for line patterns followed an approximately Lorenzian distribution, while the axial response for the dots lattices followed a Gaussian shape.

The offset in these axial responses were characterized in terms of κ, the ratio of "on" to total modulator elements, and κ' the ratio of axial response maximum to minimum.[27] Both patterns had κ = 0.041, but the line patterns had κ' = 0.18 while the dot lattices had κ' = 0.18. This implies that a line pattern will show less background than a dot pattern for equivalent κ, while giving a thicker optical section.

The axial sectioning behavior observed with the thin films was also seen qualitatively in biological specimens (Figure 9). The conventional image showed evidence of blurring from above and below the plane of best focus. In the conventional image, it is not obvious that the stained tubulin fibers are absent in the nuclei of the cells; both optical sectioning strategies showed this clearly. Removal of out-of-focus blur resulted in better visualization of the fibers and revealed the three-dimensional structure of the specimen. The section thickness was less in the dot scan than in the line scans (most apparent in the upper right hand corner of the images). The dot-scanned images, however, showed evidence of increased background in areas outside of the optical section.

3.2 Spectroscopic PAM

In the set-up of the system, two conditions degrading image quality were noted[25] that had not previously been described. Both are correctable. The first was associated

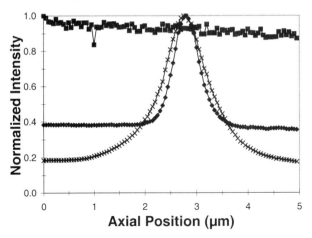

Figure 8 *Comparison of conventional illumination (■), 24 × 1 lines scans (×), and 4 × 6 dot lattice scans (♦) PAM with optical sectioning capability*

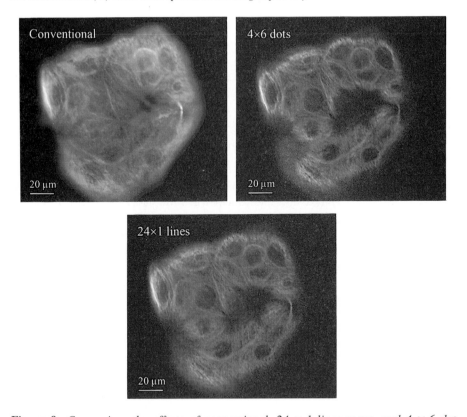

Figure 9 *Comparison the effects of conventional, 24 × 1 lines scans, and 4 × 6 dot lattice scans on images of biological specimens in an optically sectioning programmable array microscope*

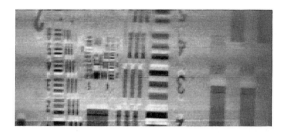
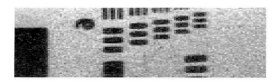

Figure 10 *Images of a 1953 USAF test target before and after correction of an LCD display error. Upper image shows two pronounced "echoes" along the transform axis. Lower image shows the same test target using the Hadamard mask aligned horizontally to the LCD axis*

with the LCD display device. When the Hadamard mask patterns were displayed with the bars oriented perpendicular to the high speed axis of the LCD address, a pair of image "echoes" of opposite sign appeared displaced along the transformed axis both above and below the primary image. The echoes disappeared when the patterns were displayed perpendicular to the low speed LCD axis (Figure 10). The second was the result of cosmic rays hitting the CCD during acquisition, resulting in multiplexing of the "spike noise" from the cosmic ray event along the transform axis (Figure 11). Such an effect can be removed if necessary with suitable algorithms but is rarely a problem in real data sets.

Observation of the human adenocarcinoma cells showed clear images in the plane of best focus, with an increasing amount of blur as the position of the focal plane was

Figure 11 *Image showing the effects of cosmic ray events on the Hadamard transform imaging. The "spike noise" signal is multiplexed along the transform axis. This effect is restricted to one image slice*

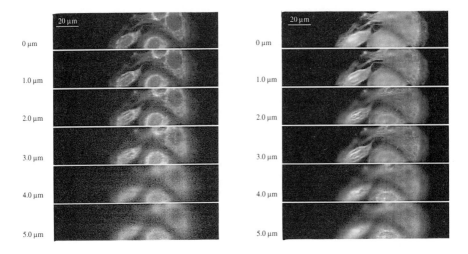

Figure 12 *Series of two-dimensional spectral images taken at 1 μm along the z-axis. Each series represents one wavelength slice in the data set. The image series on the left corresponds to approximately 630 nm (MitoTracker). The series on the right corresponds to approximately 525 nm (Oregon Green)*

adjusted (Figure 12). Both dyes were remarkably robust with respect to photobleaching, despite the fact that hundreds of images were taken while creating this four-dimensional data set (x, y, z, and wavelength).

4 CONCLUSIONS

Programmable array microscopes are unique in their ability to use structured light to accomplish a wide range of analytical tasks in a microscope. In cases where optical sectioning is needed, a PAM can best match a sectioning strategy to a particular sample. Where desired, special hybrid test patterns can be defined that allow a set of strategies to be tested simultaneously in order to select one giving the best combination of image quality, speed, and resolution. In cases where spectroscopy is desired, the Hadamard instrument is a relatively simple system capable of giving excellent spectroscopic resolution and high quality images. Combined, they form a powerful set of analytical tools for spectroscopy and imaging.

Acknowledgements

The authors wish to thank Dr. Donna Arndt-Jovin for preparation of the biological specimens.

References

1. J. B. Pawley. 'Handbook of Biological Confocal Microscopy', Plenum Press, New York, 1995.
2. M. D. Egger and M. Petrán. *Science,* 1967, **157**, 305.
3. M. Petran, M. Hadravsky, M. D. Egger, and R. Galambos. *J. Opt. Soc. Am.,* 1968, **58**, 661.
4. G. Q. Xiao, T. R. Corle, and G. S. Kino. *Appl. Phys. Lett.,* 1988, **53**, 716.
5. T. Wilson, R. Juskaitis, M. A. A. Neil, and M. Kozubek. *Opt. Lett.,* 1996 **21**, 1879.
6. R. Juskaitis, T. Wilson, M. A. A. Neil, and M. Kosubek. *Nature,* 1996, **383**, 804.
7. P. J. Verveer, Q. S. Hanley, P. W. Verbeek, L. J. van Vliet, and T. M. Jovin. *J. Microsc.,* 1998, **3**, 192.
8. Q. S. Hanley, P. J. Verveer, and T. M. Jovin. *Appl. Spectrosc.,* 1998, **52**, 783.
9. D. C. Youvan. *Nature,* 1994, **369**, 79.
10. E. R. Goldman and D. G. Youvan. *Biotechnology,* 1992, **10**, 1557.
11. J. F. Turner and P. J. Treado. *Appl. Spectrosc.,* 1996, **50**, 277.
12. H. R. Morris, C. C. Hoyt, and P. J. Treado. *Appl. Spectrosc.,* 1994, **48**, 857.
13. P. J. Treado and M. D. Morris. *Anal. Chem.,* 1989 **61**, 732A.
14. P. J. Treado and M. D. Morris. *Appl. Spectrosc.,* 1990, **44**, 1.
15. P. J. Treado, A. Govil, M. D. Morris, K. D. Sternitzke, and R. L. McCreery. *Appl. Spectrosc.,* 1990, **44**, 1270.
16. R. M. Hammaker, A. N. Mortensen, E. A. Orr, M. K. Bellamy, J. V. Paukstelis, and W. G. Fateley. *J. Mol. Struct.,* 1995, **348**, 135.
17. E. Mei, W. Gu, G. Chen, and Y. Zeng. *Fresenius' J. Anal. Chem.,* 1996, 250.
18. W. G. Fateley, R. M. Hammaker, J. V. Paukstelis, S. L. Wright, E. A. Orr, A. N. Mortensen, and K. J. Latas. *Appl. Spectrosc.,* 1993, **47**, 1464.
19. A. N. Mortensen, S. A. Dyer, R. M. Hammaker, and W. G. Fateley. *IEEE Trans. Inst. Meas.,* 1996, **45**, 394.
20. Z. Malik, R. A. Buckwald, A. Talmi, Y. Garini, and S. G. Lipson. *J. Microsc.,* 1996, **182**, 133.
21. C. J. H. Brenan and I. W. Hunter. *Appl. Opt.,* 1994, **33**, 7520.
22. R. Martinez-Zaguilan, M. W. Gurule, and R. M. Lynch. *Am. J. Physiol. - Cell Physiol.,* 1996, **39**, C1438.
23. Z. Malik, M. Dishi, and Y. Garini. *Photochem. Photobiol.,* 1996, **63**, 608-614.
24. I. J. Cohen, I. Issakov, S. Avigad, B. Stark, I. Meller, R. Zaizov, and I. Bar-Am. *The Lancet,* 1997, **350**, 1679.
25. Q. S. Hanley, P. J.Verveer, and T. M. Jovin. *Appl. Spectrosc.,* 1999, **53**, 1.
26. Q. S. Hanley, P. J. Verveer, M. J. Gemkow, and T. M. Jovin. *J. Microsc.,* 1999, in press.
27. G. S. Kino, "Intermediate Optics in Nipkow Disk Microscopes". in Ref 1, 155–165.

PRACTICAL CONSIDERATIONS FOR LN$_2$-COOLED, O-RING-SEALED, VACUUM-INSULATED DEWARS FOR OPTICAL AND IR DETECTORS

Bruce Atwood and Thomas P. O'Brien

Imaging Sciences Laboratory
Astronomy Department
The Ohio State University
Columbus, Ohio 43210-1106

1 INTRODUCTION

We consider the problem of vacuum insulation in O-ring-sealed, LN$_2$-cooled Dewars for optical and IR detectors. In addition to the obvious requirement of basic vacuum integrity, some form of cryo-sorption pump is essential to maintain effective vacuum insulation in these systems. Our experience is with cryo-sorption pumps based on zeolites and activated charcoal although other cryo-sorption materials may exist. Water has an important role in these systems, and will receive special attention even though its partial pressure is completely irrelevant at LN$_2$ temperatures.

2 BASIC PHYSICS OF CRYO-SORPTION PUMPS

The Mathcad$^®$ program given in the appendix demonstrates that it is only necessary to assume a binding energy (or heat of adsorption) for a gas onto a surface and an effective surface area to calculate the properties of a cryo-sorption pumping material such as zeolites or activated charcoal. The basic model is that the binding energy leads directly to a temperature dependent residence time, which combined with the frequency of collisions between gas molecules, and the adsorbent surface leads to a quantity of "trapped" gas. That quantity of gas varies with pressure in the same way as it would in a virtual volume. The cryo-sorption material does not pump in the sense of removing gas from the system but lowers the pressure by sharing the gas between the real and virtual volumes. This virtual-volume model is very successful in predicting the performance of cryo-sorption pumps.

The virtual volume model does not address the *rates* of adsorption and desorption. Two rates that have particularly important practical consequences are the uptake of water and the release of air. The binding energy for water in zeolite is so high that, at temperatures below about 200 C, adsorbed water is essentially permanent. When exposed to STP and normal humidity the surface of zeolites will be covered with water, saturating the zeolite, and therefore preventing other gasses from binding to the surface. It is therefore necessary to transfer the dry zeolite to the Dewar in such a fashion that it is not exposed to ambient humidity for too long a period. We have measured the rate of weight gain of previously dried zeolites held in ambient air and find that typical time constants for water absorption into canisters of a few ten of cm^3 are of the order of 10 hours. The binding energy for water on charcoal is low enough that it will quickly dry under vacuum at room temperature. The other important rate is how quickly absorbents release air when they are put in a vacuum. We have measured time constants of approximately an hour for air release of typical zeolite systems.

3 O-RING PROPERTIES

Air diffuses through the O-rings into a "sealed" Dewar at a continuous rate after equilibrium has been established. This rate is a function of the total length of O-rings in the Dewar, the type of rubber, the "squeeze" of the O-ring, and the pressure across the O-ring. The diffusion rate is nearly independent of the O-ring cross-section since the diffusion increases linearly with the area exposed to air and decreases linearly with the distance that the air has to diffuse.

The Parker O-Ring Handbook provides data on the gas permeability of air through various O-ring materials. Butyl rubber has the lowest permeability, with a value of about 2×10^{-9} Std cm^3/(cm sec). Nitrile rubber (Buna-N), more widely available, has permeability of about 5×10^{-9} Std cm^3/(cm sec). The amount the O-ring is squeezed and greased has a minor affect on the diffusion rate. The appendix shows the calculation of diffusion into any Dewar. Our instruments fall into two classes. Large instrument Dewars, approximately ½ meter in diameter by 1 meter long, with O-rings for the main vacuum shell, a window, a detector port, and a dozen other small ports, have a diffusion rate of 5×10^{-5} Std cm^3 sec^{-1}. This produces a steady rise in pressure for an un-pumped Dewar of about 5 mTorr/day. The diffusion rate for small CCD Dewars of 20 cm diameter by 35cm long with two main O-rings and several small O-rings is about 10^{-5} Std cm^3 sec^{-1} giving an un-pumped pressure rise of about 35 mTorr/day. (In practice we generally see actual diffusion rates that are two to four times lower than these theoretical rates.) Note that this calculation only includes the diffusion of air through the O-Rings and does not include pressure rises from the desorption of gases from either the walls of the Dewar nor from any cryo-sorption material that might be present. No allowance had been made for any "virtual leak", that is the slow escape of gas from a nearly sealed sub volume in the vacuum space. Trapped volumes and virtual leaks are *not* a problem since even 1×10^{-5} Std cm^3 sec^{-1} corresponds to almost *one cm^3 of STP gas per day*. So as long as the trapped volumes are only a few cm^3 at most only a few days operation are lost. *Note that these leak rates of ~ 1 STP cm^3 day^{-1} are the <u>best</u> that can be done with O-ring seals.* Furthermore, in a Dewar that is reasonably clean, one that has been wiped with acetone for example, water is the only significant species to evolve from the interior surfaces of the Dewar and water is effectively pumped as long at the Dewar is cooled with LN₂.

4 WATER IN DEWARS

Water is the principal contaminant in most vacuum systems, because all materials adsorb some water when stored in an environment with typical humidity level. Most materials adsorb a *LOT* of water. The time constant for water to desorb from room temperature surfaces in a vacuum can be very long, hundreds of hours for aluminum surfaces and the age of the universe for zeolites.

If a new "wet" Dewar is quickly pumped and cooled with LN₂ the water that desorbs from the warm surfaces will be "pumped" by the cold surfaces. LN_2 cooled zeolite can be used to pump the air that diffuses through the O-rings since the water that would otherwise saturate the zeolite will be condensed on the LN_2 cooled surfaces. When the Dewar is warmed, however, all the water that had been frozen on the LN_2 cooled surfaces will be quickly pumped by the zeolite, even though the zeolite is warm. If the Dewar is so designed it can be warmed to room temperature, backfilled with dry gas, and the zeolite changed, removing the water from the Dewar. In typical aluminum CCD Dewars with a few liters of vacuum, changing the zeolite a few times over the first few months of operation will effectively dry the Dewar. Further zeolite changes are then unnecessary as long as the Dewar is not exposed to atmospheric humidity since the air that diffuses through the O-rings is desorbed from the zeolite when warm and can be pumped from the Dewar without changing the zeolite.

The situation is quite different if charcoal is used. As with zeolite, the water will be pumped by the LN_2 cooled surfaces but when the Dewar is warmed the partial pressure of water will rise to saturation. If the Dewar is left untouched the water will be slowly re-adsorbed in the interior surfaces of the Dewar and the partial pressure of water will return to roughly the atmospheric value before the Dewar was pumped. A dramatic situation will occur if a charcoal pumped Dewar is vented to air shortly after it is warmed. The partial pressure of water will now be at saturation for room temperature. When the partial pressure of water in the introduced air is added super-saturation occurs, leading to cloud formation. The cloud will rapidly condense on the interior surfaces, including any optics or detectors present. A similar transient situation will occur with any object in the Dewar that has a thermal time constant considerably longer that the bulk of the cold surfaces. During warm-up, any water that is not pumped by zeolite will condense on these cold objects until they warm.

5 CRYO-SORBENT PROPERTIES

Linde type 4A or 5A zeolites have a capacity for adsorbing water at room temperature equal to about 15% by weight. This adsorption process is fairly rapid, resulting in totally saturated zeolite in less than 24 hours when exposed to air with a relative humidity of about 40%. This water adsorption reduces the capacity of the zeolite for cryo-pumping air to nearly zero. The zeolite must be regenerated by baking at 350 C for several hours and then stored in an airtight container prior to cryo-pumping service. Temperatures higher than 350 C will damage the zeolite.

An advantage of this large capacity for water is that zeolite acts as an extremely effective desiccant for drying Dewars. The operational disadvantage is that the zeolite must be replaced every time the interior of a Dewar is exposed to ambient conditions for more that a few tens of minutes.

Any organic solvents adsorbed by the zeolite will be driven off during the 350 C regeneration cycle required to desorb the water. Activated charcoal, such as EM Scientific #CX0640-1 activated coconut charcoal, available through Fischer Scientific, has very little capacity to adsorb water at room temperature. This lack of water capacity means that it provides no desiccation or drying function in the Dewar. Under some circumstances, it is an operational convenience that the charcoal does not require replacement or regeneration after a Dewar is cycled to room temperature and exposed to air.

The detailed computation of the cryo-pumping capacity of charcoal is given in the appendix. The heat of adsorption and surface area constants are assumed to be the same as for zeolite, as no references for these constants were found. These values are consistent with our experience with charcoal cryo-sorption systems. It is worth noting that typical virtual volumes are -10^8 cm^3 for *each gram* of adsorbent.

6 ROUGH PUMPING

As calculated in the appendix, the pressure in an evacuated Dewar without pumping will typically rise several mTorr each day. It is frequently convenient for Dewars to operate for a year without service. Thus the cryo-sorption pumping must be sufficient to lower the pressure in the Dewar from pressures of the order of one Torr to values where the thermal conductivity of the residual gas is negligible, on the order of 10^{-5} Torr. It is therefore *not* necessary to rough-pump the Dewar to very low pressures before cooling. For each few mTorr of gas left in the Dewar before cooling, only one day is lost from the maximum run length. In order to have the maximum capacity of the cryo-sorption pump it is necessary to wait for the air to desorb from the zeolite or charcoal. It is not necessary, however, to pump the Dewar continuously during this time. If fact, long periods of pumping run the risk of back streaming of oils from the pump into the Dewar with possible contamination of the detector or optics. For a CCD Dewar a very effective protocol is ten minutes of pumping, valving off the Dewar for 24 hours followed by an additional 10 minutes of pumping. For a new Dewar, or one that has never been desiccated by changing the zeolite (or by aggressive baking), the pressure will be dominated by water vapor below about 500 mTorr.

If it is desirable to pump and cool a Dewar quickly, as is the case for some detectors that require UV flooding to increase the sensitivity, a larger diameter pump line and valve will yield much greater gains than a higher performance pump. Even if the largest, and most expensive, turbo-molecular pump can maintain 10^{-7} Torr at its inlet the pressure inside the Dewar is limited by the impedance of the line and valves. Simple mechanical pumps can quickly remove the air in a Dewar and if they are not operated for long periods in the molecular flow pressures, that is below about 100 mTorr for a system with 15 mm diameter lines, contamination of the Dewar by pump fluids is all but nonexistent. Since it would be very unusual that that one is willing to wait the hundreds of hours that are required for substantially all the water to desorb from the interior, drying the Dewar by changing a zeolite canister is a much better strategy.

7 USEFUL INSTRUMENTATION

To accurately monitor the pressure in a Dewar with a cryo-sorption pump it is necessary to be able to measure pressures from 10^{-7} Torr to 10^{-3} Torr. Ionization gauges provide a very accurate, although not inexpensive, method for making measurements in this range. (Cold cathode gauges, while less expensive, require some maintenance, are much less precise, and are prone to operational problems such as difficulty starting at low pressures.) Typical measurements of interest are the initial pressure after a cool-down and monitoring of the long-term pressure rise in the Dewar due to gas diffusion into the Dewar. Regular readings of the pressure in a cold Dewar can indicate exactly when a Dewar will require pumping to remove the air that has diffused through the O-rings. We normally equip CCD Dewars with convectron gauges (available from Granville Phillips), which are reliable from atmospheric pressure to 1 mTorr, (Thermocouple gauges are less expensive but are useful only over the pressure range from 1000 to 10 mTorr.) We install both a convectron and an ionization gauge on our large instruments.

The total heat load on a vacuum-insulated Dewar can be measured by monitoring the flow rate of nitrogen boil-off gas. A simple and robust rotometer-type (ball suspended in a tapered tube) gas flow meter available from Dwyer Instruments provides adequate accuracy. Once the typical baseline boil-off flow rate has been determined any increased conductive heat load caused by pressures > a few $\times$ 10^{-5} Torr will be clearly indicated by and increased flow. Thus the flow meter can indicate a problem with the vacuum insulation long before a convectron or thermocouple gauge can measure the increased pressure. The flow meter is also a valuable tool during system development, allowing a direct measurement of the effect of any change on total power input to the Dewar and directly indicating the hold time. If pumping and cool-down are done with a fixed protocol the N_2 boil-off rate as a function of time provides an inexpensive, early, and robust indication that the Dewar system is functioning normally.

The flow meter should be calibrated in "standard cubic feet air/hour" because this fortuitously reads directly (within a few percent) in liters/day of LN_2 consumed, a very convenient unit. The directly read value can be multiplied by 2 $\times$ to convert to watts.

Dewars that have been open to air desorb a substantial amount of water during pump down. Since the mechanisms for pumping water and air are so different, it is useful to be able to measure the partial pressure of water and partial pressure of air separately. If a Dewar is rough-pumped until the pressure drops below a few hundred mTorr, a convectron or thermocouple gauge will read the total pressure reading. Adding a small amount of LN_2, just a whiff, will pump virtually all of the gaseous water and produce a rapid drop in pressure. After a minute or two the pressure will stabilize; the final value is a reasonably accurate measure of the partial pressure of air in the Dewar.

8 OPERATIONAL EXPERIENCE

The OSIRIS instrument has been in the field for six years and has used a zeolite molecular sieve canister as a pump. The canister contains about 35 grams of Linde Type 4A and 5A zeolite. The canister is mounted to the LN_2 cooled optical bench in a cold radiation environment. This pump has worked well in practice and has maintained high vacuum in the Dewar for runs of several months' duration.

The canister is accessible through an access port in the instrument, and the zeolite must be replaced every time the instrument is exposed to ambient humidity, because irreversible adsorption of water at room temperature renders zeolite ineffective for pumping air. The requirement for good access to the canister necessitated the location of the canister at the "dry" end of the LN_2 reservoir, away from the liquid LN_2. The result is that the canister is typically at temperatures as high as 85 Kelvin, which is not optimal.

The TIFKAM instrument has a Dewar similar to OSIRIS but uses an activated charcoal canister mounted inside the cold volume of the Dewar at the "wet" end of the LN_2 reservoir and is therefore at very nearly 77 Kelvin. The canister contains 70 g of coconut activated charcoal. Since charcoal doesn't irreversibly adsorb water, the charcoal does not need to be replaced every time the instrument is warmed up. The TIFKAM instrument also has an ionization gauge that allows accurate pressure measurements to be made. TIFKAM is typically pumped warm to a total pressure of about 400 mTorr, which will whiff to approximately 50 mTorr. Two hours after filling with LN_2, the pressure is typically 2×10^{-6} Torr, and after 24 hours the pressure is about 4×10^{-7} Torr.

The pressure will then rise very slowly over a long run. Periodic pressure measurements with the ionization gauge can accurately monitor the vacuum performance of the Dewar. The charcoal will maintain high vacuum ($P < 10^{-5}$ Torr) for more than a year.

Our current CCD Dewar design has a cold charcoal canister and an easy-to-change warm zeolite canister. Fresh zeolite will hold the water partial pressure below 20 mTorr and guarantee that no condensation will form. For detectors and optics that are sensitive to water (or just very expensive), good practice is to use heaters to warm them to room temperature before the LN_2 is depleted.

The appendix provides a detailed analysis of the charcoal pump performance, gas diffusion through O-rings, initial pump down pressure, and charcoal service time for the ANDICAM instrument. This information is contained within a MATHCAD 6.0 document that is available for download at our website: http://www.astronomy.ohio-state.edu/~isl/.

APPENDIX

Vacuum Performance of Adsorbent Pumped Dewar

Physical Constants

$$\text{Avagadro} := 6.02 \cdot \frac{10^{23}}{\text{mole}} \qquad M_{N2} := 28 \cdot \frac{\text{gm}}{\text{mole}} \qquad R_{kcal} := 1.99 \cdot 10^{-3} \cdot \frac{\text{kcal}}{\text{K} \cdot \text{mole}}$$

Adsorbent Properties (same values used for Zeolite or activated charcoal)

$$Q_{N2} := 4 \cdot \frac{\text{kcal}}{\text{mole}}$$ **Heat of Adsorption**

$$\text{Area}_{zeo} := 6 \cdot 10^6 \cdot \frac{\text{cm}^2}{\text{gm}}$$ **Surface area of Adsorbent**

$$m_{zeo} := 60 \cdot \text{gm}$$ **Mass of Adsorbent in one canister**

Calculate Residence Time for N2

$$\text{Temp} := 77$$ **Adsorbent Temperature**

$$\tau := 5 \cdot 10^{-14} \cdot \text{sec} \cdot e^{\left(\frac{Q_{N2}}{R_{kcal} \cdot \text{Temp} \cdot K}\right)} \qquad \tau = 0.011 \cdot \text{sec}$$ **Residence Time of gas on adsorbent**

$$C_1 := \frac{3.1 \cdot 10^{14}}{1 \cdot 10^{-4} \cdot \text{torr} \cdot .00777 \cdot \text{sec} \cdot \text{cm}^2} \qquad C_1 = 2.993 \cdot 10^{17} \cdot \frac{\text{sec}}{\text{gm} \cdot \text{cm}}$$ **collision frequency (per sq cm per torr)**

Compute Adsorbent Capacity in Several Forms

$$\text{Pressure}_{fin} := 5 \cdot 10^{-4} \cdot \text{torr}$$ **Maximum Pressure Allowed before Gas Conduction is Significant**

$$\sigma := C_1 \cdot \text{Pressure}_{fin} \cdot \tau \qquad \sigma \cdot \text{cm}^2 = 2.167 \cdot 10^{15}$$ **molecules/cm^2**

$$\text{Capacity}_{zeo} := \frac{C_1 \cdot \text{Pressure}_{fin} \cdot \tau \cdot \text{Area}_{zeo} \cdot M_{N2}}{\text{Avagadro}} \qquad \text{Capacity}_{zeo} = 0.605 \cdot \frac{\text{gm}}{\text{gm}}$$

Compute "Virtual Volume" of Adsorbent in Canister

$T = 250 \cdot K$ **"Average" Temp of Gas in dewar**

$$V_{char} := \left(\frac{Area_{zeo} \cdot \tau \cdot C_1}{Avagadro}\right) \cdot R_{gas} \cdot T \qquad V_{char} = 6.731 \cdot 10^8 \cdot \frac{cm^3}{gm}$$

$$V_{charcan} := \left(\frac{m_{zeo} \cdot Area_{zeo} \cdot \tau \cdot C_1}{Avagadro}\right) \cdot R_{gas} \cdot T \qquad V_{charcan} = 4.039 \cdot 10^{10} \cdot cm^3$$

Convert Capacity to Std cc per gram at Final Allowed Pressure

$\rho_{N2} = 1.366 \cdot kg \cdot m^{-3}$ **Density of N2 at STP**

$$Capacity\_zeolite := \frac{Capacity_{zeo}}{\rho_{N2}}$$

$Capacity\_zeolite = 442.844 \cdot \frac{cm^3}{gm}$ **Std cc of Air per gram at 77 Kelvin, P=5E-5**

Compute Net Volume of Dewar

$L_{dew} := 91 \cdot cm \qquad D_{dew} := 46 \cdot cm$

$$V_{shell} := \frac{\pi}{4} \cdot D_{dew}^2 \cdot L_{dew} \qquad V_{shell} = 151.233 \cdot liter$$

$V_{reservoir} := 30 \cdot liter$

$V_{net} := V_{shell} - V_{reservoir} \qquad V_{net} = 121.233 \cdot liter$

Compute Mass of Air Pumped during initial Pumpdown

$$P_{pump} := .01 \cdot torr \quad m_{pump} := \frac{P_{pump} \cdot V_{net} \cdot M_{N2}}{R_{gas} \cdot 293 \cdot K} \qquad m_{pump} = 1.859 \cdot 10^{-3} \cdot gm$$

$$P_{warm} := \left(\frac{Capacity\_zeolite \cdot m_{zeo}}{V_{net}}\right) \cdot 760 \cdot torr \qquad P_{warm} = 166.569 \cdot torr$$

$$Pressure_{init} := \left(\frac{P_{pump} \cdot V_{net} \cdot M_{N2}}{R_{gas} \cdot T}\right) \cdot \left(\frac{Avagadro}{C_1 \cdot \tau \cdot Area_{zeo} \cdot m_{zeo} \cdot M_{N2}}\right)$$

$Pressure_{init} = 3.002 \cdot 10^{-8} \cdot torr$ **Starting Pressure assuming equilibrium & no gradients**

Compute Diffusion Rate Into Dewar (std cc/sec)

Total Length of Nitrile O-Rings in ANDICAM Dewar

$$\text{Quantity} := \begin{pmatrix} 1 & 1 & 1 & 10 & 3 & 2 & 2 & 0 & 0 & 0 \end{pmatrix}$$

$$\text{Diameter} := \begin{bmatrix} 47 \\ 8 \\ 28 \\ 3 \\ 2 \\ 20 \\ 5.5 \\ 0 \\ 0 \\ 0 \end{bmatrix} \quad \begin{matrix} \textbf{Shell} \\ \textbf{Window} \\ \textbf{Window Plate} \\ \textbf{Connector Adapt} \\ \textbf{Connectors} \\ \textbf{Detector Ports} \\ \textbf{Detector Windows} \\ \\ \\ \end{matrix}$$

$$L := \pi \cdot \text{Quantity} \cdot \text{Diameter} \cdot \text{cm}$$

$$L_0 = 534.071 \cdot \text{cm}$$

$$F := 1.2 \cdot 10^{-8} \cdot \left(\frac{\text{cm}^3}{\text{sec}}\right) \cdot \frac{\text{cm}}{\text{cm}^2} \quad \textbf{Permeability of Nitrile O-Rings to Air}$$

$Q := .72$ **Factor for grease and % squeeze, Parker Handbook, pg A2-4**

$S := .25$ **Percent Sqeeze**

$P := 14.7$ **Pressure acroos O-ring in psi**

Formula from Parker Handbook, pg A2-4

$$\text{Diffrate} := 0.70 \cdot F \cdot \left(\frac{L_0}{\pi}\right) \cdot P \cdot Q \cdot (1 - S^2) \qquad \text{Diffrate} = 1.417 \cdot 10^{-5} \cdot \frac{\text{cm}^3}{\text{sec}}$$

$$\text{Diffrate}_{\text{Day}} := \text{Diffrate} \cdot 24 \cdot 3600 \cdot \frac{\text{sec}}{\text{day}} \qquad \text{Diffrate}_{\text{Day}} = 1.224 \cdot \frac{\text{cm}^3}{\text{day}}$$

Diffusion of air through O-rings in std cc

Compute Pressure Rise vs Time of Dewar Without Adsorbent Pumping

$$mtorr := .001 \cdot torr$$

$$Pressure\_Day := \frac{Diffrate \cdot \left(24 \cdot \frac{hr}{day}\right) \cdot \left(3600 \cdot \frac{sec}{hr}\right) \cdot 760 \cdot torr}{V_{net}}$$

$$Pressure\_Day = 7.675 \cdot \frac{mtorr}{day}$$

Compute Pressure Rise vs Time with Adsorbent Pumping

$$\frac{Diffrate_{Day} \cdot 760 \cdot torr}{V_{charcan}} = 2.304 \cdot 10^{-8} \cdot \frac{torr}{day}$$

Compute Dewar Hold Time (78 K Adsorbent, Maximum Pressure = 5*10-5 Torr)

$$Hold\_time := \frac{m_{zeo} \cdot Capacity_{zeo} - Capacity_{zeoinit}}{Diffrate \cdot \rho_{N2}}$$

$$Hold\_time = 2.17 \cdot 10^{4} \cdot day$$

PHARMACEUTICAL REACTION MONITORING BY RAMAN SPECTROSCOPY

Jonathan G. Shackman, Jeffrey H. Giles, and M. Bonner Denton

Department of Chemistry
The University of Arizona
Tucson, AZ 85721

1 ABSTRACT

In recent years, major advances in lasers, optical systems, and detectors have led to the development of Raman spectrometers capable of sensitive and rapid analysis of a variety of samples. These spectrometers have many highly desirable advantages, such as low detection limits and the ability to analyze aqueous solutions. Raman spectroscopy has also demonstrated its ability to perform quantitative analysis of multicomponent mixtures. The application of Raman spectroscopy to qualitative and quantitative monitoring of pharmaceutically important reactions was investigated. Two model systems were chosen to demonstrate the effectiveness of Raman: the esterification of ethyl alcohol with acetic acid to form ethyl acetate, a pharmaceutical flavoring agent; and the acetylation of salicylic acid to form acetylsalicylic acid (aspirin). It was demonstrated that quantitative reaction monitoring by Raman spectroscopy is possible. The method was also sensitive to impurities within the reaction.

2 INTRODUCTION

Raman spectroscopy has a number of useful applications and several distinct benefits as an analytical method. It is a scattering technique, and as such, it does not require sample preparation, an extremely desirable advantage. In addition, broad linear dynamic ranges have been obtained experimentally for many samples.[1] As water has a weak Raman signal, aqueous solutions can be analyzed directly without requiring extractions from solution; Raman analysis is therefore suitable for many organic reactions or analyses within biological systems. A large percentage of chemical compounds are Raman active; almost every molecular species produces Raman spectra except metals and simple salts such as sodium chloride.[2] Raman can be considered a complementary technique to infrared (IR) spectroscopy, as Raman vibrations are governed by different selection rules. Molecules are Raman active when there is a change in polarizability during the vibration, while IR-active transitions must demonstrate a change in the dipole moment. For molecules containing a center of symmetry, the mutual exclusion principle shows that transitions that are Raman active are not in the IR, and vice versa. Noncentrosymmetric molecules may contain transitions that are active in both; the intensities, however, may be quite different for a given vibration. Homopolar diatomic gases are active only in Raman, and Raman has a higher sensitivity to carbon-carbon

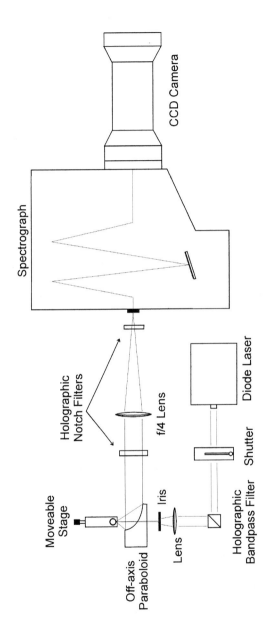

Figure 1 *Schematic diagram of the Raman spectrometer used for model reaction monitoring*

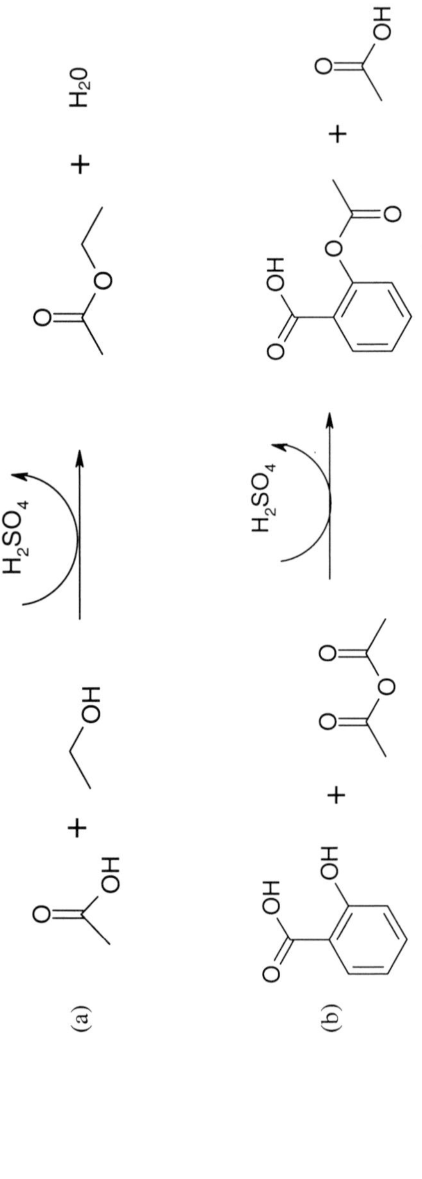

Figure 2 *The model reactions monitored: (a) acid-catalyzed synthesis of ethyl acetate from ethanol and acetic acid; and (b) acetylation of salicylic acid by acetic anhydride to form aspirin*

bonds, making it a powerful technique for organic molecule characterizations.[1] The relatively few overtone and combination bands present in Raman, as compared to IR, yield much simpler spectra. This often allows mixtures to be analyzed. Since Raman is a scattering technique, Raman lacks the cell path-length issues that frequently arise in IR methods, due to the high molar absorptivities requiring shorter path lengths, making solids and liquids readily analyzable by Raman. Currently, spectral libraries are being compiled by companies such as Aldrich and are being used to facilitate the ease of qualitative analysis.

Raman has several instrumental advantages in addition to these fundamental considerations. Raman spectrometers can be coupled with fiber-optic probes for remote measurements, which allows for more flexible sampling techniques and for the ability to insert probes into hazardous reaction mixtures.[3] Relatively inexpensive and easily replaceable glass optics and sampling containers can also be used. Solid-state lasers and rugged components make Raman ideal for industrial applications, as they require little maintenance and enable the performance of analyses right on the factory floor. As no additional solvents are required, such as in chromatography or in extractions, this nondestructive technique produces no additional waste that would necessitate disposal. In another recently discovered instrumental advantage, high-performance detectors such as CCD cameras can be employed to provide a flexible, sensitive, and rapid method of detection. When high-rejection volume phase holographic filters are used, detection of the weak Raman signal is possible. Overall, the modern Raman instrument is relatively inexpensive to build, maintain, and operate.

The use of Raman analysis to monitor reactions is appearing more frequently in the literature. Such varied examples as stream composition analysis, nuclear waste characterization, and polymerization reaction have been investigated using Raman spectroscopy, with impressive results.[2,4,5] Raman's sensitivity to carbon-carbon bonding has allowed for monitoring of organic isomerization reactions.[6] Also, as aqueous solutions can be analyzed, more and more biological applications are being seen, such as monitoring of peptide synthesis and enzymatic reactions.[7,8]

Rapid, nondestructive assessment of the composition and completeness of reactions is of great interest to the pharmaceutical industry. Current analysis in this field often involves time-consuming and costly off-line methods. Typically, samples are removed from the reaction stream and subjected to invasive techniques such as extractions, digestions, and chromatography.[9,10] These techniques often require up to an hour to complete, during which time the reaction stream continues to flow. If a problem is discovered during analysis, the components of the stream that have passed from the time of sample collection to when the problem was discovered must be discarded. Raman, with its combined advantages, appears to be an ideal solution for on-line, real time analysis.

It was demonstrated that Raman could indeed be a useful technique for analyzing pharmaceutical reactions. The goals were to develop a fast, nondestructive, and versatile method for on-line reaction monitoring, with total analysis time of less than 10 s employing a nondestructive method that involved no sample preparation. In addition to qualitative analysis of the principle components, impurities within the reactions and reaction completeness were investigated. Experimental models employed ethyl acetate and acetylsalicylic acid (aspirin) synthesis. Ethyl acetate is used as a pharmaceutical flavoring aid and as an extraction solvent in the pharmaceutical and food industries.[11] Both reactants and products are liquid, with water a byproduct. Aspirin is commonly

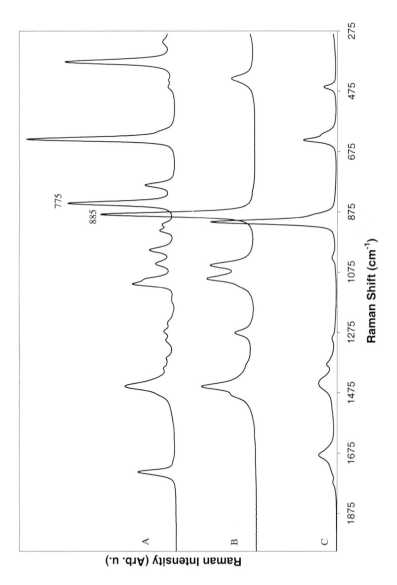

Figure 3 *Raman spectra of (A) ethyl acetate, (B) ethanol, and (C) acetic acid. The peaks monitored for ethanol and ethyl acetate are marked at 885 and 775 cm^{-1}, respectively*

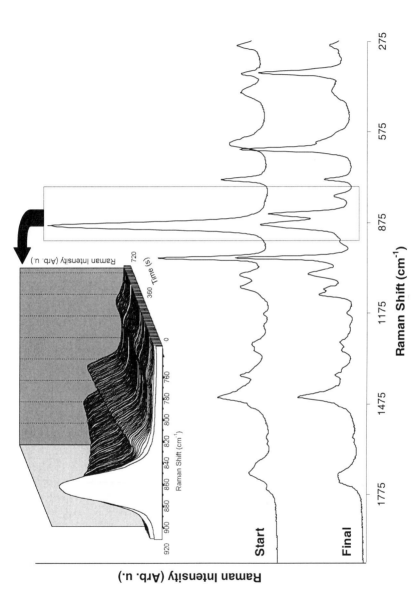

Figure 4 *Raman spectra taken at the start and end of the 1:1 ethyl acetate synthesis. Insert plots Raman bands of ethanol and ethyl acetate acquired throughout the duration of reaction*

used as an analgesic. The reaction mixture consists of salicylic acid suspended in acetic anhydride, with the aspirin product also being a solid suspended in solution.

3 EXPERIMENTAL

A schematic representation of the custom-built Raman instrument used is given in Figure 1. The excitation source was a 785-nm broad stripe diode laser that was wavelength stabilized with an external cavity. To remove any extemporaneous wavelength emission, the beam was passed through a holographic bandpass filter (Kaiser Optical Systems, Inc.). The laser was then focused onto the sample through an f/4 lens. Scattered light was collected via a gold-coated, off-axis parabolic mirror and was then passed through two high-rejection holographic notch filters (Kaiser Optical Systems, Inc.) to remove Rayleigh scatter. The Raman scatter was focused by an f/4 lens onto the spectrograph's slit (Oriel Instruments, model MS257). Throughout both experimental setups, the entrance slit was set at 100 μm using a 600 groove/nm grating blazed at 1 μm. The detector was a 1024×1024 Tektronix CCD, a backside-illuminated array interfaced with a Photometrics (Tucson, AZ, USA) CC200 CCD controller. A neon lamp and neat cyclohexane were used to calibrate the wavelength. A custom LabVIEW (National Instruments) software program was used for instrumental control and data acquisition. All reactions were run in quartz cuvettes.

Chemicals were reagent grade and were obtained from either Mallinckrodt or Aldrich. Each spectrum was obtained with a 5-s integration time. Total time resolution for consecutively-collected spectra during a reaction was 8 s. As both syntheses were acid-catalyzed (Figure 2a and b),[12,13] monitoring began once the catalyst was added. All reactions were unheated and were run at 25°C; reactions were continually mixed mechanically. Reactions were run at various concentrations of starting reactants, as given in Tables 1a and b.

4 RESULTS AND DISCUSSION

4.1 Ethyl Acetate Synthesis

First, the two reactants, ethanol and acetic acid, and the product, ethyl acetate, were individually sampled to determine whether isolated spectral peaks would be present or whether any components fluoresced. The acid catalyst and the water byproduct were neglected, the former because of its low concentration and the latter because of its small Raman cross-section. The spectra of the pure components appear in Figure 3. Two isolated and nearly adjacent peaks appeared to arise from ethanol (850-915 cm^{-1}) and

Table 1a *Ethyl acetate synthesis starting quantities*

Run	Ethanol (mL)	Acetic Acid (mL)	Sulfuric Acid (μL)
1:1	1.00	1.00	100
1:2	0.50	1.00	100
3:2	1.50	1.00	100
3:4	0.75	1.00	100
5:4	1.25	1.00	100

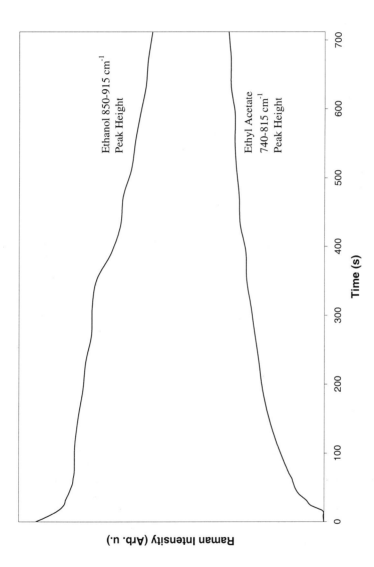

Figure 5 *Plot of the calculated peak heights of the ethanol and ethyl acetate bands throughout the duration of the 1:1 ester synthesis reaction*

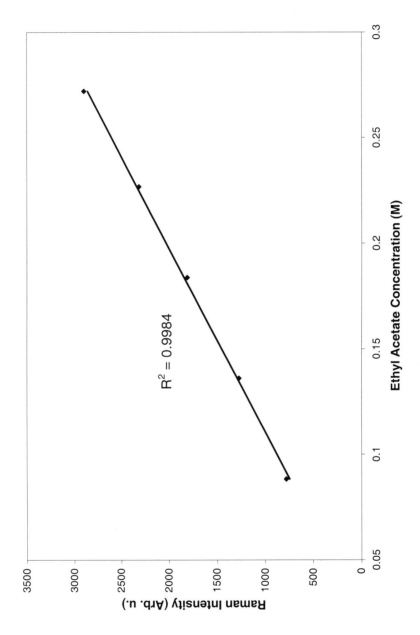

Figure 6 *Plot of ethyl acetate peak heights versus concentrations as determined from a standard calibration set*

Table 1b *Aspirin synthesis starting quantities*

Run	Salicylic Acid (g)	Acetic Anhydride (mL)	Sulfuric Acid (μL)
1	0.1380	0.300	100
2	0.1035	0.300	100
3	0.0690	0.300	100
4	0.0345	0.300	100

ethyl acetate (740-815 cm^{-1}). Their lack of overlap from other peaks made quantitative analysis easier, as their heights were directly related to their concentration, making more complex, nonlinear data analysis unnecessary. As the reactant and product peaks were nearly adjacent, the effects of any nonlinear trends within the baseline were limited.

Timed acquisition began when the H_2SO_4 catalyst was added. The reactions were run for 12 min and spectra were collected every 8 s. The complete spectra from the beginning and the end of the reaction are given in Figure 4. The peaks utilized for quantitation of the reactants and products for the duration of the reaction are shown in the insert. While there was some slight baseline shifting throughout a run, the peak heights were calculated with respect to the baseline, so the effects of the shift were minimal. The calculated ethanol and ethyl acetate heights over time are graphed in Figure 5. Attenuation of the ethanol peak height corresponds to its consumption in the reaction, with concomitant growth of the ethyl acetate peak height. Trends manifesting in the ethanol peak heights appear to have a correlation in the ethyl acetate peak heights, such as that which appeared at 400 s. Similar results were obtained with all starting concentrations; only the rate of formation was affected.

Standard mixtures consisting of the calculated amounts of ethanol, acetic acid, and ethyl acetate corresponding to various reaction yields were made, excluding the acid catalyst (Table 2), and were analyzed. The peak heights corresponding to ethyl acetate were correlated with the concentration, creating a calibration curve upon which the yield of a reaction could be determined (Figure 6).

The effect of a realistic impurity was also examined. Reactions were compared using pure ethanol versus that which used ethanol with a 5% butanol impurity. Butanol reacts with acetic acid in a manner similar to ethanol, yielding butyl acetate as the product.[12] Spectra from the end of the impure run show the formation of unwanted product. Figure 7 shows the final spectra of the reactions run with both pure ethanol and ethanol with a butanol impurity. The spectrum clearly indicates the formation of the butyl acetate impurity.

Table 2 *Standard mixtures of ethyl acetate for calibration.*

Ethanol (mL)	Acetic Acid (mL)	Ethyl Acetate (M)
0.25	0.25	0.272
0.50	0.50	0.227
0.75	0.75	0.184
0.80	0.80	0.136
0.90	0.80	0.088

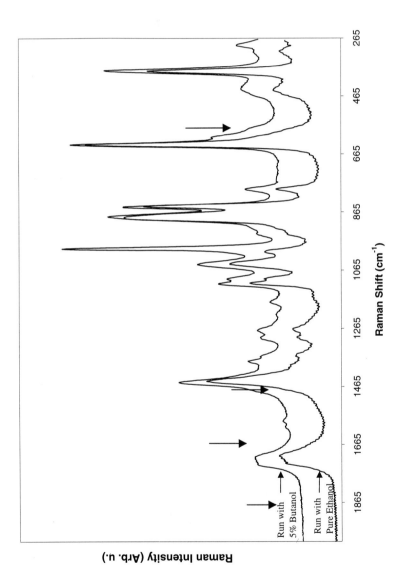

Figure 7 *Spectra comparing ethyl acetate synthesis run with pure ethanol with that containing a butanol impurity. Bands denoted by arrows correspond to the formation of butyl acetate*

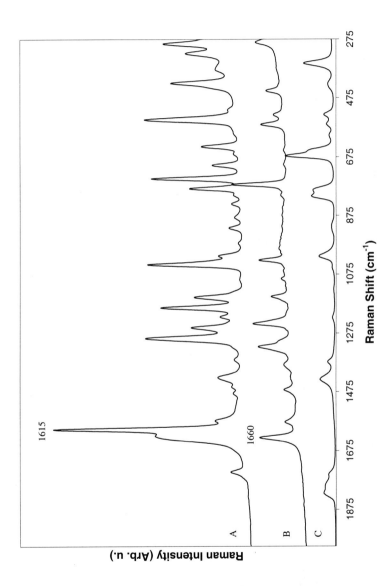

Figure 8 *Raman spectra of (A) acetylsalicylic acid (aspirin), (B) salicylic acid, and (C) acetic anhydride. The salicylic acid and aspirin peaks monitored are marked at 1660 and 1615 cm⁻¹, respectively*

4.2 Aspirin Synthesis

The method for monitoring the synthesis of aspirin followed that of the ethyl acetate model. Spectra of the individual components are given in Figure 8. Again, baseline-resolved and adjacent peaks were chosen for ease of quantitative analysis. The salicylic reactant peak occurred from 1630 to 1690 cm^{-1}, and the aspirin peak lay from 1600 to 1630 cm^{-1}. Complete spectra from the beginning and conclusion of the reaction are provided in Figure 9. The reactions were monitored for 6 min each. Timed acquisition began once the acid catalyst was added, with spectra collected every 8 s. The peaks corresponding to the reactant and product for an entire reaction are given in the insert of Figure 9. Slight baseline shifts are also present in this model; however, the actual peak heights follow a uniform trend of degradation of the reactant with increase in the product peak height (Figure 10). Results were similar at the other varying concentrations of starting solutions.

5 CONCLUSIONS

Raman spectroscopy has demonstrated its applicability to on-line monitoring. It has several distinct advantages over other methods, most notably its speed and cost efficiency. Aqueous solutions and those wherein water is a byproduct, such as the ethyl acetate synthesis, are well suited to Raman, as opposed to other spectroscopic techniques such as IR. The total time resolution between acquired spectra of only 8 s far surpasses methods requiring timely extractions or lengthy chromatographic analysis. A Raman instrument on a factory floor can quickly and easily detect the composition of reaction streams and can flag operators should a drop in yield or an impurity arise in the stream. Even a small and chemically similar impurity such butyl acetate in ethyl acetate can readily be observed. Faster acquisitions can be achieved by either decreasing the integration time or by reading a smaller spectral window from the CCD. Obviously, with more complicated systems, more rigorous data analysis techniques may be required, as peaks become more overlapped and nonlinear changes occur within spectra.[1] However, the sharp, distinct peaks observed with Raman make this a viable next step. Raman also has the advantage of utilizing fiber optic probes, which enables more direct analysis, even in hazardous streams. Most notably, the lack of necessity for sample preparation should make Raman reaction monitoring a powerful analytical technique of the future.

Acknowledgements

The authors gratefully acknowledge the financial support of The University of Arizona Undergraduate Biology Research Program and the Howard Hughes Medical Institute.

References

1. F. Adar, R. Geiger, and J. Noonan, *Appl. Spectrosc. Reviews*, 1997, **32**, 45.
2. T. J. Vickers and C. K. Mann, *SPIE*, 1995, **2367**, 219.
3. T. J. Vickers and C. K. Mann, *SPIE*, 1992, **1637**, 62.
4. E. D. Lipp and R. L. Grosse, *Appl. Spectrosc.*, 1998, **52**, 42.
5. A. Brookes, J. M. Dyke, P. J. Hendra, and A. Strawn, *Spectrochimca Acta A*, 1997, **53**, 2303.

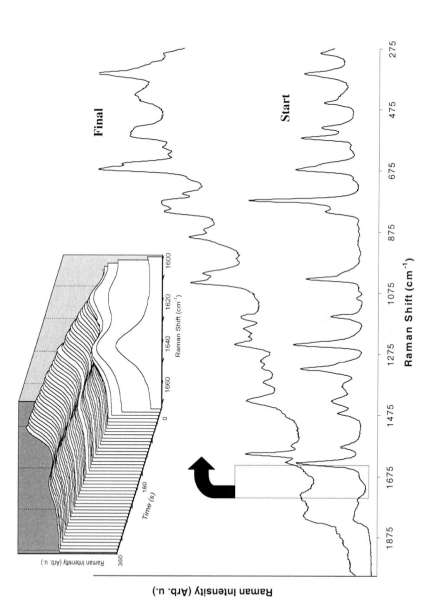

Figure 9 *Raman spectra taken at the beginning and end of the aspirin reaction (Run 1). Insert plots Raman bands of salicylic acid and aspirin acquired throughout the duration of the synthesis*

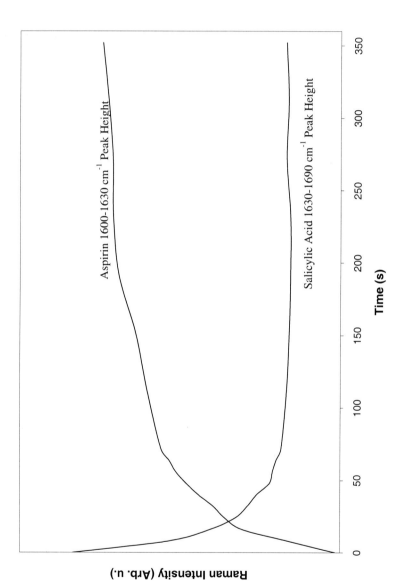

Figure 10 *Plot of the calculated peak heights of the salicylic acid and aspirin bands throughout the duration of the Run 1 synthesis reaction*

6. M. L. A. Temperini, M. R. dos Santos, and V. R. Paoli Monteiro, *Spectrochimica Acta A*, 1995, **51**, 1517.
7. J. Ryttersgaard, B. Due Larsen, A. Holm, D. H. Christensen, and O. Faurskov Nielsen, *Spectrochimica Acta A*, 1997, **53**, 91.
8. X. Dou and Y. Ozaki, *Appl. Spectrosc.*, 1998, **52**, 815.
9. The United States Pharmacopeia, *The United States Pharmacopeia: The National Formulary 22nd Ed.,* United States Pharmacopeial Convention, Inc., Rockville, MD, 1990.
10. E. A. Cutmore and P. W. Skett, *Spectrochimica Acta A*, 1993, **49**, 809.
11. *The Merck Index: 12th Ed.*, ed. S. Budavari, Merck & Co., Inc., Whitehouse Station, NJ, 1996.
12. E. Sarlo, P. Svoronos, and P. Kulas, *J. Chem. Ed.*, 1990, **67**, 796.
13. J. W. Lehman, *Operational Organic Chemistry: A Laboratory Course*, Allyn and Bacon, Inc., Boston, 1981.

Subject Index